KT-147-149

Education for a Sustainable Future

A Paradigm of Hope for the 21st Century

INNOVATIONS IN SCIENCE EDUCATION AND TECHNOLOGY

Series Editor:
Karen C. Cohen, Harvard University, Cambridge, Massachusetts

Are Schools Really Like This?: Factors Affecting Teacher Attitude toward School Improvement
J. Gary Lilyquist

Education for a Sustainable Future: A Paradigm of Hope for the 21st Century
Edited by Keith A. Wheeler and Anne Perraca Bijur

The Hidden Curriculum—Faculty-Made Tests in Science
Part 1: Lower-Division Courses
Part 2: Upper-Division Courses
Sheila Tobias and Jacqueline Raphael

Internet Links for Science Education: Student–Scientist Partnerships
Edited by Karen C. Cohen

Science, Technology, and Society: A Sourcebook on Research and Practice
Edited by David D. Kumar and Daryl E. Chubin

Time for Science Education
Michael R. Matthews

Web-Teaching: A Guide to Designing Interactive Teaching for the World Wide Web
David W. Brooks

A Continuation Order Plan is available for this series. A continuation order will bring delivery of each new volume immediately upon publication. Volumes are billed only upon actual shipment. For further information please contact the publisher.

Education for a Sustainable Future

A Paradigm of Hope for the 21st Century

Edited by

Keith A. Wheeler and Anne Perraca Bijur

The Center for a Sustainable Future
a division of the Concord Consortium
Shelburne, Vermont and Concord, Massachusetts

Kluwer Academic/Plenum Publishers
New York, Boston, Dordrecht, London, Moscow

ISBN: 0-306-46420-9

233 Spring Street, New York, New York 10013

http://www.wkap.nl

10 9 8 7 6 5 4 3 2 1

A C.I.P. record for this book is available from the Library of Congress

Printed in the United States of America

To Courtney and Ross Wheeler and to Oliver Bijur
May their lives and the lives of their children be filled with
the Hope, Passion, and the Opportunity
to create a truly Sustainable Future!

Contributors

Richard Benjamin, Superintendent, Cobb County Public Schools, Marietta, GA

Jack Byrne, Program Director, Center for a Sustainable Future, Shelburne, VT

Mark DiMaggio, High School teacher of Earth and Environmental Sciences, Paso Robles High School, Paso Robles, CA

John Fien, Faculty of Environmental Sciences, Griffith University, Brisbane, Australia

Cynthia Georgeson, Director Worldwide Communications, S.C. Johnson Corp., Racine, WI

Wendy Goldstein, Head Environmental Education & Communication, IUCN, Gland, Switzerland

Susan Hanes, Director of the Metro Atlanta P-16 Council, Georgia State University, Atlanta, GA

Frits Heselink, Chairman, Commission on Education and Communication, IUCN, Gland, Switzerland

Rupert Maclean, Chief, Asia-Pacific Centre of Educational Innovation for Development UNESCO, Principal Regional Office for Asia and the Pacific, Bangkok, Thailand

Jean MacGregor, Director, National Learning Communities Dissemination Project, The Washington Center for Improving the Quality of Undergraduate Education at The Evergreen State College, Olympia, WA

M. Patricia Morse, Professor of Biology, University of Washington, Seattle, WA

Lynn Mortenson, Director Education and Outreach, U.S. Global Change Research Program, Washington, D.C.

Mary Paden, Resource Center Director, Academy for Educational Development, Washington, D.C.

Anne Perraca Bijur, Coordinator, Building Education for a Sustainable Society, Shelburne Farms, Shelburne, VT

Jean Perras, Executive Director, Learning for a Sustainable Future, Ottawa, Ontario, Canada

Larry Peterson, Director, Florida Sustainable Communities Network, Florida A&M, Tallahassee, FL

Alan Sandler, Executive Director, Architectural Foundation of San Francisco, San Francisco, CA

Michael Schneider, Dean, Washington Program—Maxwell School of Business, Syracuse University, Washington, D.C.

Bradley Smith, Dean, Huxley College, Western Washington Univ. Bellingham, WA

Keith Wheeler, Director, Center for a Sustainable Future, Shelburne, VT

Lori Wingerter, Coordinator, Community Impact Team, General Motors Corporation, Detroit, MI

Foreword

WHY EDUCATE FOR SUSTAINABILITY?

Ask most people—especially young people—what the future looks like to them. They will paint for you a rosy picture of their individual lives, their future homes, their careers and sometimes even the lives of their children. Then they will paint for you a depleted, conflicted and weary picture of the planet as a whole. Do any of us note the incongruence? We are becoming increasingly aware that the very things we need, the things we adore most in this world are the very things we are undermining through our individual and collective behaviors. We can feel the disconnect between the consequences of our actions and our values, but we do not necessarily understand it or know how to reconcile it.

Living with behavior that feels, and in fact is, inconsistent with our values is a recipe for anguish. We have disconnected so drastically from our sense of place that ironically we poison our water, our air, and our food while at the same time we work hard to secure a healthy and meaningful future for ourselves and our children. This just doesn't make sense.

Hope is the union of a vision with the capacity to realize it. Education for sustainability helps us to create optimistic visions and develops in us the capacity for actions that are consistent with those visions. It makes sense for us to cultivate hope in our children for themselves and for the world around them. What would be the point of being educated for a future for which we have no hope?

Another reason to educate for sustainability is that it is the ethical thing to do. Aldo Leopold wrote: "All ethics so far evolved rest upon a single premise: that the individual is a member of a community of interdependent parts. His instincts prompt him to compete for his place in the community but his ethics prompt him also to co-operate (perhaps in order that there may be a place to compete for)." Simply put, it is the right thing to do.

Finally those of us who have already been provoked to educate for sustainability continue to do so because it is the most ambitious, creative, elegant, innovative, out of the box, exhilarating, difficult, frustrating, comprehensive, enlightening, rewarding, holistic, ego-boosting, and humbling work there is to do. And we would not miss it for the world.

Whoever you are and whatever compels you to move, to grow, and to change, this book is an invitation to you to learn, and to educate for a sustainable future.

Jamie P. Cloud, Founder and Director
The Sustainability Education Center
New York, NY

Preface to the Series

The mandate to expand and improve science education is an educational imperative and an enormous challenge. Implementing change, however, is very complicated given that science as well as science education is dynamic, continually incorporating new ideas, practices, and procedures; takes place in varying contexts; and must deal with amazingly rapid technological advances. Lacking clear paths for improvement, we can and should learn from the results of all types of science education, traditional as well as experimental. Successful reform of science education requires careful orchestration of a number of factors which take into account technological developments, cognitive development, societal impacts and relationships, organizational issues, impacts of standards and assessment, teacher preparation and enhancement, as well as advances in the scientific disciplines themselves. Understanding and dealing with such a complex mission is the focus of this book series. Each book in this series deals in depth with one or more of these factors, these potential factors for understanding, creating and sustaining effective science education improvement and reform.

In 1992, a multidisciplinary forum was launched for sharing the perspectives and research findings of the widest possible community of people involved in addressing the challenge. Those who had something to share regarding impacts on science education were invited to contribute. This forum was the *Journal of Science Education and Technology.* Since the inception of the journal, many articles have highlighted relevant themes and topics and expanded the context of understanding to include historical, current, and future perspectives in an increasingly global context. Recurring topics and themes have emerged as foci requiring expanded treatment and presentation. This book series, "Innovations in Science Education and Technology" is the result.

It is a privilege to be able to continue to elucidate and effect improvement and reform in science education by providing this in-depth forum for the works of others. The series brings focus and understanding to efforts worldwide, helping readers to understand, to incorporate, and to utilize what we know, what we are learning, and what we are inventing technologically to advance the mission of science education reform worldwide.

Karen C. Cohen
Cambridge, Massachusetts

Contents

Introduction

Keith Wheeler

Educating today's youth to have the requisite knowledge, skills, and values to shape their lives and the world around them successfully is our greatest challenge. In this book, we have chosen the concept of sustainability as the framework to refocus our education system to achieve the noble goal of a better future for all life. We wish to engage current generations to adequately care for their economic, social, and environmental well being, while ensuring that subsequent generations can achieve an even better quality of life.

Many people worldwide are struggling to articulate what sustainability means and how to characterize it. This is tricky because it is a *way* of thinking as much as what you are thinking about. Three systems are often talked about most when describing sustainable development—environmental, economic, or social. These systems have many subsystems, which require the notion of **systems thinking** to better understand. There are numerous possible **interconnections** between these systems. We also have a nearly limitless number of ways that humans can observe, interpret, and communicate their view of these interactions. This gives rise to the notion that sustainability needs to be viewed through **multiple perspectives**.

The education community has not found it easy to shift readily from specific, discrete educational topics to a more integrated systems approach. Our vision of education for a sustainable future is focused on how to get beyond reduction and analysis—with which we are most comfortable—to the synthesis and integration of what we know and can know. Likewise, a convergence and integration of environmental, economic, and social

Keith Wheeler, Director, Center for a Sustainable Future, Shelburne VT USA.

systems is core to our work. This is why and where educating about sustainability becomes increasingly complex. We often try to come at sustainability from *one* direction based on our predisposition (from an environmental viewpoint, an economic viewpoint, or a social issue or quality of life viewpoint). To engage successfully in sustainable development we must train ourselves to think holistically. Education about sustainability in essence is about learning to make and understand the connections and interactions between these three complex systems.

The challenge we have as educators is to:

- First, be aware of the necessity to cultivate systems thinking, interconnections, and multiple perspectives as students learn about sustainability.
- Second, to gain some skill in understanding and applying this thinking and seeing, to our own lives and communities.
- Third, to understand the "ah ha!" factor—*when* and *how* did we first see the world through new eyes of sustainability. What was the event or process that led to this transformation? And to be able to reflect on and communicate this to others who have had a similar experience.
- Fourth, based on the synthesis of these shared experiences, how can we teach or construct experiences for our students so that they achieve the "ah ha!" and have the self-awareness to understand and reflect on its meaning.

Our task is not an easy one, but it is crucial if we are going to create an educational system worthy of our hopes. By seeing the future through the lens of sustainability, we will be able to find ways that assure we will make life better for all for a long time to come.

SUSTAINABILITY FROM FIVE PERSPECTIVES

We have divided the concept of sustainability into several pieces for students to learn about with the thesis that they will then be able to induct a wholesome understanding of sustainability from these parts. Traditionally, this topic would be divided along the lines of the environment, the economy, and social equity. It is our feeling that this would continue to promulgate sustainability as three discreet fields and make the integration to the whole more difficult. Alternatively, we envision five themes that cut across the themes of the environment, the economy, and social quality of life for all.

1. *Thinking about and Affecting the Future.* Sustainability is bound in the future. We are making some choices today about how we live in order to affect a better future for our children and grandchildren. Research has shown that most students have a very short future view, ranging from days to weeks depending on their economic background and other factors. Sustainability requires that we take a long view, often many generations into the future, in order to be better prepared as the future unfolds. Our challenge is to kindle in students a strong desire to look into the full range of possible futures, understand the systems that affect the future, provide them with techniques to plan for the future, and empower them to affect the future positively both for themselves and for their communities. Their future view needs also to be seen through the lenses of the environment, the economy, and an equitable society.
2. *Designing Sustainable Communities.* The concept and practice of design is very important in helping us move into a more sustainable world. We have an opportunity to shape our lives in many ways. The idea of students having the knowledge, skills, and values to pursue quality, equitable, and creative design is essential in solving the sustainable conundrum. Good design needs to account for environmental and social attributes and is bound by economic reality. We focus on community because it is the largest organizing scale that is appropriate for engaging a high proportion of students and citizens in creating sustainability. It readily brings together the environmental, social and economic systems. It is where I live. It is where I derive my "sense of place." And it is a good point from where we can move from reality to abstraction in the environmental, economic, and social systems that are key to sustainable development.
3. *Stewardship of Natural Resources.* We selected the concept of stewardship because of its long history in the environmental and conservation movement in the US and around the world. This is a result of the thinking and writings of Aldo Leopold, Gerald Olson, Gifford Pinchot, Hugh Hammond Bennett, Rachel Carson, and others who have shaped our concept of environmental stewardship. Stewardship, meaning "responsible care," is a cornerstone of acting in a sustainable fashion. We might have chosen to focus on economics, or social responsibility, but research by Gerald Lieberman and others has shown

the great affinity that students have for learning about the natural world. We want to leverage this affinity into an easy entry point into understanding environmental systems.

4. *Sustainable Economics*. Economics has been the traditional study of the flow and movement of capital (wealth) throughout our society. This has personal, family, community, national, and international implications. We have chosen the EINS^2 factors as a means of describing the sustainable thinking we have conceived.

   Traditional economics generally deals with the acquisition or distribution of monetary capital; our view of sustainable capital includes this from of traditional economic capital (E). It also includes the concept of intellectual capital (I), whereby, in a knowledge-based economy the greatest asset an individual can have is a rich well-developed intellectual capacity (the same for a company, community, or nation). The third type of capital is natural capital (N), or the quantity and quality of highly productive natural ecosystems supporting rich biodiversity, including humankind. The fourth type of capital is social capital (S) which focuses on quality of life issues for all citizens and which is key to human dignity and satisfaction. The fifth and final capital is spiritual capital (S), which is the underpinning of the "human spirit". This capital has no specific organized religious connotation, although many individuals, cultures and communities often build and renew the human spirit through places of worship. It is the avenue for human self-renewal.

   It is through the enhancement of these five capitals that we see the concept of sustainability being built in our society. A sustainable society or sustainable development is when all these capitals are in balance.

5. *Globalization*. Finally, understanding the process of globalization is essential if we are to begin to think about how our aggregate actions cause change beyond our communities and ourselves. Globalization is a force that is occurring visibly around us and has a tendency to make us as individuals feel insignificant and powerless. Many members of the "sustainability community" focus on thinking and acting locally. It is where we as individuals have the greatest opportunity for impact. We concur with this, but feel that as educators we must see and help our students see the whole.

> There are many case studies showing that most communities and civilizations that practiced "island thinking" have not sustained themselves for the long haul. External forces, information, disease, etc. have adversely affected cultures because they did not build contingencies. It is our hope that we can enable students to understand how forces external to their community or individual lives affect the outcome of their desired futures, and how their actions have an influence beyond their immediate sphere.

Globalization in and of itself will create a whole new set of system dynamics between our three primary areas of focus (environment, economy, and social equity). These areas are going to be constantly unfolding over the next several decades. We must prepare our students to recognize and interpret the signs of globalization and how to become more resilient to its impacts.

These five areas of study each have the economic, environmental, and social equity viewpoint embedded within, as well as the opportunity for understanding and applying the underlying systems, interconnections, and multiple perspectives necessary to create an "ah ha!" within the learner.

After experiencing five themes a student should be able to compose a full view of sustainability leading to enlightened personal action at the individual and community level that takes into account:

- a deep understanding of complex environmental, economic, and social systems;
- recognition of the importance of the interconnectedness of these systems in a sustainable world;
- and respect for diversity of "points of view" and interpretations of complex issues from cultural, racial, religious, ethnic, regional, and intergenerational perspectives.

The knowledge and skills embodied in sustainability will engage lifelong learners with the knowledge, skills, and values necessary to be active participants in a sustainable globalizing society in the 21st century. It is truly a paradigm of hope and empowerment for all.

The following chapters represent a variety of thinking and value that is being placed on the concept of educating for sustainability from many perspectives. It is author's hope that the readers will reflect on the fullness and richness that this framework offers, and begin to look to integrate the concepts of sustainability into their educational constructs.

CHAPTER 1

Education for Sustainability and Environmental Education

Mary Paden

I was sitting in a conference room at the President Hotel in Gaborone, Botswana at a circle of conference tables with about 40 environmental education representatives of 12 Southern African countries. We were in day two of a workshop to critique a report proposing environmental education policy guidelines to the countries belonging to the Southern African Development Community (SADC). We had been moving nicely through the agenda when someone suggested that we replace "Environmental Education" in the title and throughout the report with "Education for Sustainability."

Then ensued the same lengthy discussion that I have heard among environmental educators in Guadalajara, Mexico at the Iberoamerican conference of environmental educators; Vancouver, British Columbia at the North American Association for Environmental Education conference; and Washington, D.C. at meetings of the President's Council on Sustainable Development Working Group on Education. The discussion goes like this:

> One person proposes the change in terminology, education for sustainability is the new term. It is broader than environmental education. UNESCO (the United Nations Educational Cultural and Scientific Organization, the organization

Mary Paden, Resource Center Director, Academy for Educational Development, Washington D.C.

> assigned to follow up on educational matters from the Earth Summit conference in 1992) is asking everyone to move towards using the term and concept of Education for Sustainability.
>
> At first everyone agrees. Everyone supports sustainability.
>
> Then someone—usually an environmental educator from the environment ministry of a national government—says "The people I work with are just getting used to the idea of environmental education. If I start talking about education for sustainability they will ask, 'How is that different? Why do you keep changing terms? You are confusing me.'"
>
> Then another participant notes, "Environmental Education is a field. It has graduate schools and professional journals. After 25 years of building up this much infrastructure, why should we change the name of the field? There is no field of education for sustainability. They don't have any graduate schools."
>
> Then a sustainability proponent points out, "We educators have to change and grow as new ideas develop. The Earth Summit conference changed the way we look at the environment. It showed that we must consider the economy and equity issues."
>
> "But we have always considered the economic, political and social context," the E.E. proponents say. That is what makes E.E. different from Conservation Education or Nature Education. E.E. and Education for Sustainability are the same and we are just hurting ourselves with funders and politicians by trying to switch names at this point."

I was one of those arguing in favor of using the term "sustainability" at the Garborone meeting, when the leader of a prominent E.E. Center nudged me and pointed to a passage in the document we were reviewing that read, "Environmental education recognizes the importance of viewing the environment *within the context of human influences, incorporating an examination of economics, culture, political structure and social equity as well as natural processes and systems*." So there, he was saying—they are the same.

"No, no," I scribbled back in a note passed across a gap in the diplomatically arranged tables. "That statement shows exactly the difference! E.E. examines the environment against the *backdrop* or *in the context of* economics, social equity, etc., but the values are mainly around the environment. Education for sustainability demands that environment, equity and economics be considered as a whole and its values involve promoting all three together."

As an appendix to our conference documents were the Tblisi Principles—the founding declaration of environmental education formed by a small council of very wise people in 1977 in Tblisi, Georgia. The first principle said that an environmental education program should "consider the environment in its totality—natural and built, technological and social (economic, political, cultural-historical, moral and aesthetic)." In this founding document there was a unification of natural, built, social, political environments that is achieved in many, but not all classrooms. However, the Tblisi principles focus only on *values* related to "environmental sensitivity" and on the involvement of learners in "planning their learning process"—participation/equity value applied specifically to students, but not extrapolated to society in general.

The concern about poverty and social equity—the realization that protecting the natural environment in a world of growing misery and poverty did not claim the moral high ground—was driven home by the people from developing countries who attended the Brundtland Commission hearings in the 1980s and the Earth Summit in 1992. People from developing countries taught environmentalists and environmental educators that some level of economic prosperity and social equity for all was essential for a sustainable future and a livable world. For sustainability proponents of the late 1990s, the challenge is to get beyond the new and exciting management/technical approaches to energy efficiency and managed ecosystems to grapple with the more complex issue of how to promote all three E's together.

Previously environmentalists and environmental educators have presented the three Es as a series of necessary tradeoffs. Environmental preservation can be had, but only at a loss to the economy. Social justice had to be traded off against economic prosperity and environmental purity. Jobs versus Environment. Development versus Preservation. Justice versus Growth. You can find this line of reasoning in any environmental science textbook for the 1970s or 1980s. But was it ever true?

It may have been true in the short-term heat of debate. It may have been true from the daily newspaper deadline perspective or the paying-next-month's-bills perspective. Yes there did seem to be that type of conflict to the logger who lost his job because logging was stopped to protect a rare owl or to the steel plant workers whose plant was closed rather than upgraded to meet pollution standards. But listening to the people from impoverished countries who testified to the Brundtland commission and the Earth Summit, we found out that in the longer term—over a lifetime perhaps—there is no conflict. The environment will never be protected as long as poverty forces people to use the Earth's resources to survive. A campaign to stop economic growth will never rally the masses. The only sane course is to look beyond the short-term conflict to the longer term need for a common vision and for as much cooperation as possible.

We can learn some of the Earth Summit's lessons in our own center cities. A few weeks before the Gaborone workshop I was in Detroit, which was preparing to host the National Town Meeting for a Sustainable America in May, 1999. For the previous two years I had served on the Education Working Group of the President's Council for Sustainable Development which published a set of recommendations for education for sustainability and a set of indicators to measure progress. Detroit was one of several communities that had taken on the challenge to develop a local set of goals and indicators. At this meeting a group of educators and high school students were presenting their set of indicators on education for sustainability for Detroit. Most of the indicators had to do with making sure students attended classes, that the quality of teaching was improved, and that the school environment was safe from violence.

When we (people for the PCSD group) were asked to comment on the presentation, someone said, "Well this is interesting, but you don't have anything in here about sustainability." Indeed neither the term nor the concept had been used. (Nor was there anything about the natural environment.)

The head of the Detroit group, the president of a local college, replied, "You know, we read the material in your report and we didn't really understand what you meant, so our group just made a list of the things they needed to improve their school environment."

Back to Tblisi: "consider the environment in its totality—natural and built, technological and social (economic, political, cultural-historical, moral and aesthetic)." Don't we have to begin with the school environment? Doesn't it have to be safe, healthy and functional? Doesn't it have to promote learning and creative exploration? E.E. and Education for Sustainability can't be a frill. Sustainability education, at least, must consider the core issues of education as part of its mandate. The Detroit presentation was about sustainability. We sustainability experts just didn't get it.

A figure started taking shape in my mind. I was trying to visualize the relation of E.E. and Education for Sustainability. After a page of doodling it turned into a star. Below is my E.E. star.

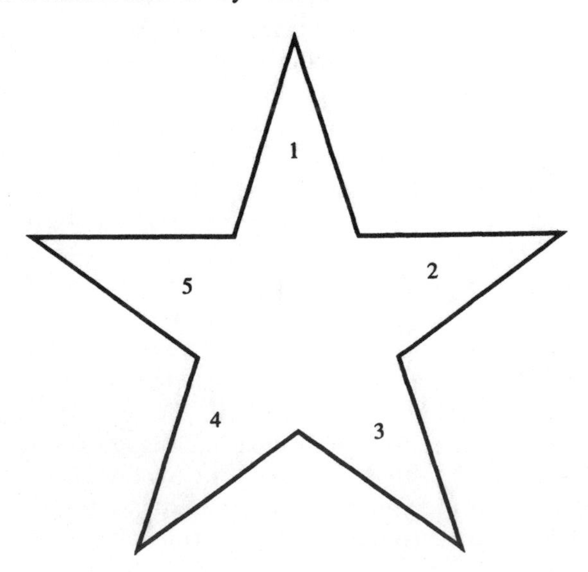

Figure 2 below shows the Sustainability Star with a few additions and changes (in capital letters) made in the past two decades.

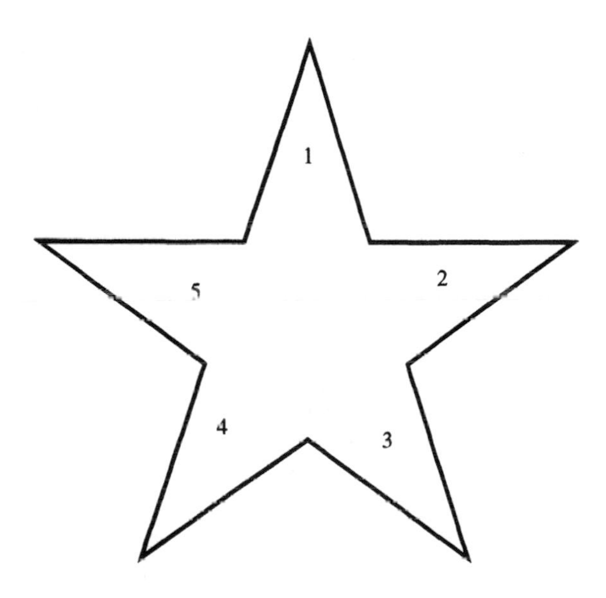

Figure 2

1. Content, environment (natural and built) in context of social, political, economic. Focus on local to global issues and solutions. ADD CONTENT ABOUT ECONOMICS AND EQUITY. MORE CONTENT ON TECHNOLOGY AS A SOLUTION, AND ON BUSINESS AS A MORE THAN IN THE 70'S.
2. Life long learning process (formal and informal).
3. Methods: interdisciplinary, learning centered, experiential, inquiry based uses broad array of interactive techniques. STRESS ON PARTNERSHIPS WITH GOVERNMENT, BUSINESS, NGO'S, EDUCATORS, EMPHASIS ON SYSTEMS THINKING.
4. Focus on citizen action skills.
5. Values: Environment sensitivity. ADD POSITIVE VALUES ABOUT SOCIAL EQUITY AND ECONOMIC PROSPERITY. INSEPARABILITY OF THE 3 E's.

Figure 1

1. Content, knowledge of natural systems, understanding of social and political systems that influence natural systems. Positive attitude toward the natural environment.
2. Context: formal (schools) and nonformal (zoos, museums, courses, etc. for adults and children).
3. Methods: interdisciplinary, learner centered, experiential, inquiry based, interactive. Emphasis on bringing people to natural environment.
4. Action: Learn appropriate skills for decisionmaking and citizen action. Practice environmentally sound behaviors.
5. Values: Environmental protection in social and economic context.

Note that nothing is taken away from environmental education. On the other hand you don't have to be an environmental educator to educate about sustainability. Educators from the fields of population education, energy education, urban planning, and development education embraced the idea of education for sustainability, although they may know little about the natural environment.

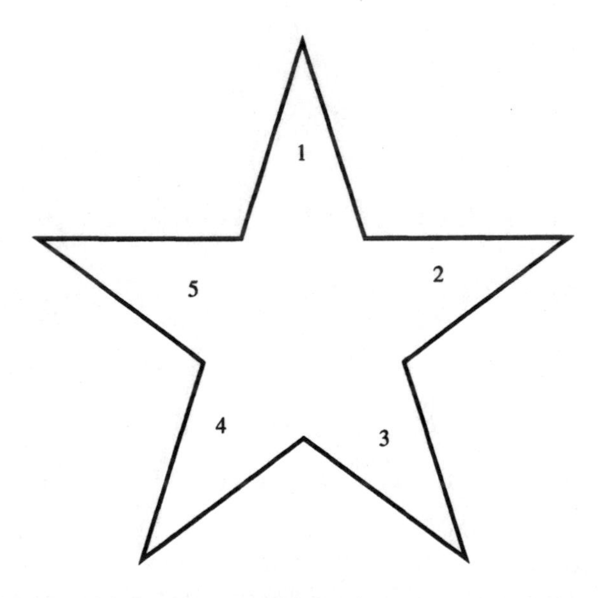

Figure 3

1. Content, environment (natural and built) in context of social, political, economic. Focus on local to global issues and solutions. ADD CONTENT ABOUT ECONOMICS AND EQUITY. MORE CONTENT ON TECHNOLOGY AS A SOLUTION, AND ON BUSINESS AS A MORE THAN IN THE 70'S.
2. Life long learning process (formal and informal).
3. Methods: interdisciplinary, learning centered, experiential, inquiry based uses broad array of interactive techniques. STRESS ON PARTNERSHIPS WITH GOVERNMENT, BUSINESS, NGO'S, EDUCATORS, EMPHASIS ON SYSTEMS THINKING.
4. Focus on citizen action skills.
5. Values: Environment sensitivity. ADD POSITIVE VALUES ABOUT SOCIAL EQUITY AND ECONOMIC PROSPERITY. INSEPARABILITY OF THE 3 E's.
6. Basic literacy numeracy. Safe schools. Basic job skills.

Figure 3 integrates the experience of the Detroit education indicators group into the star. They are at the very center. Without a center of basic literacy, numeracy, and love of learning, the arms of the star are unlikely to extend very far. For the K-12 level this center required a good foundation in the basics. For high school and community colleges it means basics plus job skills, without which the goals of social equity and economic prosperity will not be met. Clearly, the countries (and cities) that have managed to produce an educated and skilled workforce are the ones that rise from poverty. Development literature is just as clear as any immigrant grandmother that education is the key to a better life—and a more sustainable future.

CHAPTER 2

Global Change Education
Educational Resources for Sustainability

Lynn L. Mortensen

Look around. What do you see? Everything you see is a result of someone's decision, and some of the things you see are the result of your own decisions. This fundamental right and privilege, to make our own decisions, is often unnoticed, taken for granted, or even forgotten. Yet one of the most empowering gifts education in this country can bring to individuals young and old is the recognition that we each have a role in the shaping of our lives, our homes, our workplaces, our communities, our world. And that is why education is so vital—not indoctrination—but empowerment—through access to information that allows us to make those decisions with the best information available. So how can we make better decisions, in the face of uncertainty, to ensure the carrying capacity of the Earth, quality of life for all living things, a bright future for generations that follow? This is essentially the goal of the international research community engaged in the investigation of Earth as a system.

A paradigm shift is a fundamentally altered way of viewing things, which changes one's whole perspective and allows us to understand and

Lynn L. Mortensen, Director Education and Outreach, U.S. Global Change Research Program, Washington, D.C.

value things in new ways. I remember the headlines in 1957 when the Soviet Union successfully launched the first spacecraft, Sputnik. I was in 6th grade and I can still recall the sense of fear and competitiveness in the country when this event occurred. Would this mean Soviet domination of the world? The U.S. response was an intensified space program and a heightened priority for science education. The underlying values were fear and competitiveness. Then a fundamental shift occurred when we received the first pictures of ourselves from space. The political boundaries were indiscernible and the "blue marble" was suspended against a backdrop of deep black space. All of a sudden, we became aware of our vulnerability and of our interconnectedness. The Earth was one large intricate interconnected system, a perspective familiar to indigenous cultures. Values such as co-operation and collaboration emerged as essential—bumper stickers and slogans such as "Protect our Home" with the picture of the "blue marble" and "We're all in this Together" reinforced these values.

TRANSITION

The United Nations Education, Scientific, and Cultural Organization (UNESCO) articulated the role of values in moving toward a sustainable future. ". . . Sustainable development is widely understood to involve the natural sciences and economics, but it is even more fundamentally concerned with culture: with the values people hold and how they perceive their relations with others. It responds to an imperative need to imagine a new basis for relationships among peoples and with the habitat that sustains human life. (1997, UNESCO)

The public shift in thinking and values toward a more holistic approach was concurrent with emerging international scientific cooperation in attempting to understand Earth as a system. Charles David Keeling began taking measurements of carbon dioxide in the atmosphere in 1957 from Mauna Loa, Hawaii. Those data revealed increasing atmospheric CO_2 at the planetary scale. The National Academy of Sciences issued reports from panels and committees organized under leading scientists such as Roger Revelle and Jule Charney in the 1970's and early 1980's hypothesizing that the burning of fossil fuels could enhance the greenhouse effect.

Scientists recognized the need to share data and jointly pursue research investigations to develop holistic, integrated, interdisciplinary descriptions of how the system the Earth worked, and how each element interacted with the others. A statement by the National Research Council in 1983 is illustrative. "If we believed that the Earth was a constant system in which the atmosphere, biosphere, oceans, and lithosphere were unconnected parts, can it be a system if there are unconnected parts? Then the traditional scientific fields that study these areas could all proceed at their own pace treating each other's findings as fixed boundary conditions. However, not only is the Earth changing even as we seek to understand it—in ways that involve the interplay of land and sea, of oceans, air, and biosphere—we cannot even presume that global change will be uniform in space and steady in time . . . Needed to resolve this complex of change and interplay are coordinated efforts between adjacent scientific disciplines and programs of synoptic observations focused on common, interrelated problems that affect the Earth as a whole." (1983a, National Research Council)

Continued studies of potential global change impacts led to calls for international assessments of the implications of a changing climate. In response, the United Nations Intergovernmental Panel on Climate Change (IPCC) was formed in 1988. Hundreds of scientists representing over 50 countries assumed responsibility for conducting an international assessment on climate change and its consequences under the auspices of the IPCC. By the mid-90's, IPCC projected that increased burning of fossil fuels may result in a rapid increase of atmospheric CO_2 with potential global average temperature increases of 1.5 to 4.5 °C by 2100 producing worldwide climate changes. Subsequent periodic assessments continue to be conducted, representing scientific consensus around issues of global magnitude. These assessments have become the primary reference for the state of the science of global change.

So what exactly is global change? How is it defined? "Global change is a term intended to encompass the full range of global issues and interactions concerning natural and human-induced changes in the Earth's environment. The U.S. Global Change Research Act of 1990 defines global change as 'changes in the global environment (including alterations in climate, land productivity, oceans or other water resources, atmospheric chemistry, and ecological systems) that may alter the capacity of the Earth to sustain life.' Global change issues include understanding and predicting the causes and impacts of, and potential responses to: long-term climate change and greenhouse warming; changes in atmospheric ozone and ultraviolet (UV) radiation; and natural climate fluctuations over seasonal to interannual time periods. Other related global issues include desertification, deforestation, land use management, and preservation of ecosystems and biodiversity." (1995, *Our Changing Planet*, p. 2)

An international effort to understand the Earth system and investigate these issues is not without challenges. For example, how do you measure temperature worldwide? How will you get consistent, accurate readings over time at different places in the ocean—in heavily forested land—over the poles—How do you account for the heat island effect in cities? How do you know if this is as it always has been or if there is a change? How do you measure sea level rise? Deep ocean circulation? How do you measure the melting of glaciers? Changes in precipitation? Land use changes? Vegetation? Species diversity? How do the interactions among elements in the system, e.g., oceans, land, air, interact with and affect each other? How do you measure ozone depletion? Is it possible to predict future changes based on changes occurring now? How can we determine the impact of human activity? The IPCC international scientific community was mobilized and produced the first international assessment report in 1990, a supplementary report in 1992, and a second assessment in 1995. The 1995 IPCC Second Assessment Report concludes that the evidence now available "...points toward a discernible human influence on global climate." (1995 IPCC, p. 439) Yet the complexities of the Earth system are so immense that even though we have enormous data sets, we still have much to understand and uncertainties about how the system will react under changing conditions.

This uncertainty is the place where science is vulnerable to criticism and to calls to "wait and see until we know for sure" that come from various interest groups. Yet, the very nature of science as a discipline requires a constant state of uncertainty because as one question is answered, several more become apparent and need to be investigated. One of the main objectives of science education is to increase the comfort level with scientific uncertainty and inspire the curiosity to ask new questions.

When it comes to sustainability, the "ability to sustain," it makes sense for many reasons to take a conservative approach to every decision we make. Scientific uncertainty aside, it no longer serves us well to "keep up with the Jones's" but rather replace that ethic with knowing when "enough is enough." And knowing when enough is enough is directly dependent on our understanding of how things work, how they are connected, and the consequences of actions to other parts of the system. The concept of sustainability allows us to simultaneously consider economic, environmental, and social aspects of our actions. The science of global change contributes to an increased understanding of what those economic, environmental, and social factors are, how they are changing, and how they may impact the system in the future. Education opens a window to the world of discovery as we continue to learn more about global change and how to move forward on a sustainable path.

INTERNATIONAL AGREEMENTS

The emerging scientific understanding of how the Earth system works has led to international policy development. The 1987 Montreal Protocol to the

Vienna Ozone Convention and subsequent amendments leading to a complete phaseout of CFC's (chloroflourocarbons, as found, for example, in freon for cooling air conditioners and refrigerators) by 1996 is an outstanding example of international cooperation on a global environmental issue. As scientific understanding of the atmosphere grows, international policy response has resulted in a leveling off of chlorine concentrations in the stratosphere and an expectation that these levels will decrease in the near future. This example illustrates the interplay between science and policy, and that human activity can and does influence the Earth system.

Agenda 21, signed by 178 heads of state at the Rio Earth Summit in 1992, is the international consensus on the interlocking issues facing humanity and the steps that must be taken by the global community to achieve a sustainable future. In particular, Agenda 21 sought the integration of development needs with those of ecologically sustainable management of the environment. The UN Commission on Sustainable Development (CSD) monitors progress on Agenda 21. (1977, IUCN)

UNESCO was designated by the UNCSD as the taskmaster for Chapter 36 in Agenda 21, the chapter concerned with education. In the UNESCO work plan, the following issues were identified as critical for the world community to address:

- the rapid growth of the world's population and its changing distribution;
- the persistence of widespread poverty;
- the growing pressures placed on the environment by the worldwide spread of industry and the use of new and more intensive forms of agriculture;
- the continuing denial of democracy, violation of human rights and the rise of ethnic and religious conflicts and violence, gender inequity; and
- the very notion of 'development' itself, what it has come to mean and how it is measured.

"It is therefore, not only necessary to deal with the problems but even more essential to get our thinking right: to see interrelations among [these problems] and recognize the fundamental need to develop a new perspective rooted in the values of sustainability. It is this need which makes education the key to creating a sustainable future." (1997, UNESCO, p. 8)

The recent meeting of 160 nations in Kyoto, Japan in December 1997 led to the Kyoto Protocol to the 1992 United Nations Framework Convention on Climate Change.

Article 2 Section 1(a)(i–iv) reads:
Each party . . . in order to promote sustainable development, shall:

(a) Implement and/or further elaborate policies and measures in accordance with its national circumstances, such as:
   (i) Enhancement of energy efficiency in relevant sectors of the national economy;
   (ii) Protection and enhancement of sinks and reservoirs of greenhouse gases not controlled by the Montreal Protocol, taking into account its commitments under relevant international environmental agreements; promotion of sustainable forest management practices, afforestation and reforestation;
   (iii) Promotion of sustainable forms of agriculture in light of climate change considerations;
   (iv) Promotion, research, development and increased use of new and renewable forms of energy, of carbon dioxide sequestration technologies and of advanced and innovative environmentally sound technologies . . .

The Kyoto Protocol calls for reduction in greenhouse gas emissions to at least 5% below 1990 levels by 2008–12 (1997, Kyoto Protocol, Article 3). These and other international agreements demonstrate the link between science and policy. The link between policy and action is dependent in large part on education.

GLOBAL CHANGE EDUCATION AND SUSTAINABILITY

"Educators readily understand that this is not an additional, imposed curriculum burden, but a perspective which permeates all disciplines and creates a context for integrated and creative learning . . . Education can create and strengthen awareness that human beings are an integral part of the environment and that their actions determine the permanence or the extinction of life on the planet . . . Amongst the skills suggested for the future are complex problem solving skills, critical reflection and creative thinking, synthesis, how to manage projects, interpersonal expression, information management, consensus building and conflict resolution . . . Educating people is a dynamic interaction between the educator and the student in which both learn and are affected by the interaction, although not in the same ways as for quality and intensity of outcomes." (1997, IUCN)

The voluntary national standards in various subject areas issued in the last five years provide a framework for integrating concepts of sustainability and issues of global change into curricula. Educators are utilizing these standards to give credibility and support for incorporating these concepts in multiple disciplines. As cautioned in 1993 by the Federal Coordinating Council for Science, Engineering, and Technology, "... the actual level of public understanding about the basic scientific concepts that explain and offer solutions to environmental threats such as those posed by global change and wetlands destruction remains limited. Without sufficient knowledge and training, the public may wish to respond to an environmental challenge, but may not be able to do so effectively because they lack sufficient scientific understanding of the problem." (1993, Federal Coordinating Council for Science, Engineering, and Technology)

The national science standards and the national geography standards readily show an acceptance of the need for students of all ages to understand the complexities of our global environment. For example, the National Research Council National Science Education Standards, published in 1994 by the National Academy of Sciences, suggests several standards that link to global change:

> **For grades K-4:**
> **Science in personal and social perspectives**: personal health, characteristics and changes in populations, types of resources, and changes in environments. **Unifying concepts and processes**: change, constancy, and measurement.
>
> **For grades 5–8:**
> **Life science**: populations and ecosystems, diversity and adaptations of organisms; **Science in personal and social perspectives**: personal health, populations, resources, and environments, natural hazards, risks and benefits; **Unifying concepts and processes**: evidence, models, and explanations; change, constancy and measurement; evolution and equilibrium.
>
> **For grades 9–12:**
> **Physical science**: chemical reactions, forces and motions, conservation of energy and increase in disorder, interactions of energy and matter; **Life science**: the interdependence of organisms; matter, energy, and organization in living systems; **Earth and space science**: energy in the Earth system; origin and evolution of the Earth system; **Science in personal and social perspectives**: personal and community health, population growth, natural resources, environmental quality, natural and human-induced hazards, science and technology in local, national, and global challenges; **History and nature of science**: nature of scientific knowledge; historical perspectives; **Unifying concepts and processes**: evidence, models, and explanation; change, constancy and measurement.

Similarly, the National Geography Standards, published in 1994 in *Geography for Life*, include standards that link to global change, such as:

> **The world in spatial terms**: how to use maps, how to analyze the spatial organization of people, places, and environments on Earth's surface;
>
> **Places and regions**: the physical and human characteristics of places, how culture and experience influence people's perceptions of places and regions;
>
> **Physical systems**: physical processes that shape the patterns of Earth's surface, the characteristics and spatial distribution of ecosystems on Earth's surface;

Human systems: characteristics, distribution, and migration of human populations on Earth's surface; patterns and networks of economic interdependence on Earth's surface; processes, patterns and functions of human settlement;

Environment and society: how human actions modify the physical environment, how physical systems affect human systems, the changes that occur in the meaning, use, distribution, and importance of resources.

While only two of the subject area national standards are described here, it is essential that multiple disciplines be engaged in the many faceted aspects of global change, including the human dimensions. As stated in the Thessaloniki Declaration at an international conference in 1997, "All subject areas, including the humanities and the social sciences, need to address issues related to environmental and sustainable development. Addressing sustainability requires a holistic, interdisciplinary approach which brings together the different disciplines and institutions while retaining their distinct identities." (1997, Thessaloniki Declaration)

To enable educators to better understand global change science and sustainability and integrate these concepts into formal curricula and informal education programs, the U.S. Global Change Research Program has conducted an aggressive national education campaign since 1993. (1997, Climate Action Report) Components include Systemic Change, Products and Resources, Professional Development, and Student Support.

SYSTEMIC CHANGE

Recognizing that education is a state responsibility; the Global Change Education program focuses on building state capacity to integrate global change concepts on a system-wide basis. The National Science Foundation State Systemic Initiative provided a model for facilitating statewide integration of global change concepts. The USGCRP state team initiative is designed to develop a national literacy and teaching capability in global change education among U.S. educators and community leaders through statewide systemic approaches. Results of research in science and social science are communicated through integration in statewide core curricula; professional and association meetings at regional and national levels; and programs conducted in museums, science centers, and community groups.

Organized in state teams, professional educators partner with non-government organizations, state government officials, and business, to design and implement state action plans. The five member state teams representing 47 states have gathered at a national global change education conference in Washington, D.C. in 1994; at seven simultaneous regional conferences incorporating two national videoconferences to develop state action plans in 1995; and at state-level planning and organizing meetings in 1996. The National Science Foundation (NSF) offered 10 planning grants at $10,000 each for a total of $100,000; the National Aeronautics and Space Administration (NASA) offered $100,000, $2,000 for each state team to help support strategic planning meetings in their state; and NASA and the Environmental Protection Agency (EPA) awarded $750,000, $30,000 each to 25 states in a competitive grant process for implementation of state action plans. This two-year grant cycle ends in 1998. Intentions are to reconvene state teams before 2000 to share advances in the integration of global change issues into educational programs and to disseminate new information and materials on regional environmental, social, and economic impacts of global change.

EDUCATION PRODUCTS/RESOURCES

The Global Change Education Resource Guide is a multi-media set of materials including a videotape of scientists providing a scientific overview of global change; a CD-ROM, full color overhead transparencies and slides with scripts; fact sheets from the Information Unit on Climate Change in Geneva, Switzerland and the World Meteorological Organization; published articles for general audiences; teacher developed and pilot tested classroom activities; and a resource bibliography. Topics include natural variability, greenhouse effect, sea level rise, ozone depletion, ecosystem response, and decision-making under scientific uncertainty. A total of 26,500 sets of materials were distributed directly to formal and informal educators nation-wide through the state teams; the informal global change educator network; the GLOBE (Global Learning and Observations to Benefit the Environment) program; the JASON project, an advanced technology electronic field investigation; the National Science Teachers Association (NSTA); and internationally through the Department of State (DoS) overseas posts and Peace Corps. Additional sets of materials were duplicated for further distribution at train-the-trainer education workshops on global change held throughout the country. Development of the Resource Guide was a collaboration among several of the USGCRP federal agencies. The compilation and design concept was supported by the National Oceanic and Atmospheric Administration (NOAA) and the multi-media guide was produced and distributed by the U.S. Department of Agriculture (USDA), NASA, the Department of Energy (DoE), the Department of Defense (DoD), DoS, and EPA.

The Department of Interior (DoI) US Geological Survey (USGS) produced 20,000 copies of the Global Change Teacher Packet with a poster and set of classroom activities, subsequently reformatting it for internet distribution. The DoI/USGS/EROS data laboratory provided copies of *the Historical Landsat Data Comparisons: Illustrations of the Earth's Changing Surface* publication for distribution to the state team network and informal educator network. The DoI/Park Service produced *Interpreting Global Change: A National Park Service Communicator's Handbook* for distribution to all park interpreters in the National Park system. NOAA produced a series of monographs entitled *Reports to the Nation*, including "The Climate System," "Our Ozone Shield," "El Nino and Climate Prediction," and "Our Changing Climate." NOAA and the National Safety Council also produced and distributed a journalist's guide to global change, *Reporting on Climate Change: Understanding the Science.*

PROFESSIONAL DEVELOPMENT

Informal education (including education programs in museums, science centers, nature centers, youth programs, extension programs, community civic programs, eco-tourism, and adult continuing education) is a vehicle for providing information and resources to a wide variety of interested constituents. An intensive four-year train-the-trainer program with over 7,000 informal educators was conducted through the NOAA/Sea Grant Program, USDA, and NSF from 1993–1996. Program elements included five national video conference training sessions with educators gathered at seven regional workshops, materials distribution, and train-the-trainer dissemination through follow-up workshops, informational workshops, professional conference presentations, published articles/newsletters, and materials development.

For the 25th anniversary of Earth Day in April of 1995, the USGCRP produced a highly successful three-hour live videoconference, broadcast nation-wide from Washington, D.C. The program was entitled "A Gathering for the Earth" and featured Native Elders in a round table discussion, live performances from the Ellipse in front of the White House, success stories of sustainable practices in communities and businesses, and students in Moscow and D.C. interviewing the astronaut and cosmonauts aboard the MIR Space Station. The global change content was organized around four concepts, air, water, land, and living beings. Over 1500 known down link sites participated in the program, with interactive call-in question and answer periods. Participants represented school systems, museums, colleges, libraries, communities, and businesses. The videotape of the program was reproduced and provided for subsequent educational use to all down link sites.

STUDENT SUPPORT

Hands-on experiential learning is the pedagogical basis for student monitoring programs such as the federally supported GLOBE program. Students participate as scientists in monitoring and reporting environmental conditions through science protocols. This program also incorporates technology to link students worldwide in an active network for information sharing and data acquisition. Similar private sector student monitoring programs such as GREEN (Global Rivers Education Network), the River Watch Network, and the TERC Global Learning Laboratory can be accessed along with the GLOBE program through web sites listed at the end of this chapter.

A global change category in the international science fair was supported for two years by the USGCRP. Participation in the science fair is accessible to every school in the country. Environmental issues continue to dominate the category selection of students participating in the science fair, and global change is a popular subcategory of science projects.

LESSONS LEARNED

Global change is complex and hard to comprehend at both spatial and temporal scales. The spatial scale of the globe is so large that it is difficult to grasp and even more difficult to understand how an individual may influence such a vast entity. Temporal or time scales of a century or more are so far in the future that they are meaningless for everyday decision-making. This is essentially the lesson learned from five years of national programs on global change education. The program focused on understanding the global context on century time scales, reflecting the status of scientific knowledge. Yet people wanted to know what this meant for them in their part of the world. Computer models are reasonably accurate on the global scale yet lack finer resolution at the regional scale. It is further argued in the UNESCO work plan that "... complex realities are difficult to communicate in simple messages. Yet, attempts to simplify what, by its very nature is not simple, may result in further confusion and misunderstandings and, ultimately, in lack of credibility ... The basic dictum of pedagogy is to begin where the learner is. This is also good advice for the communication specialist. Start with problems that people feel and understand at the local level. That is both valuable knowledge in itself and, if need be, a basis for moving on to more complex and global understandings." (1997, UNESCO, p. 19).

Opinion polls indicating the public concern, particularly of young people, show environment as a priority. In the 1992 *Health of the Planet* survey of individuals in 24 countries, Dunlap, Gallup and Gallup report high level concern about environmental degradation both in this country and globally. The Hart 1994 public opinion poll conducted for the National Wildlife Federation indicated concern about human health and the environment, and ranked the environment as one of the most serious problems society will face in the year 2000. Concern about health and the environment was also reported from a study of 16 countries by Harris, Tarrance, and Lake in 1989. The 1994 survey for the National Environmental Education and Training Foundation ranked the environment as the problem people are most concerned about and want to improve (51 percent), second only to AIDS (64 percent). (1996, EPA Report) Students from disadvantaged areas also held concern for the environment as significant (43 percent) behind AIDS, kidnapping, guns, neighborhood crime and violence, and the economy (Roper, 1994). Bloom analyzed public opinion data on environmental issues collected in two international surveys. The data reveal substantial concern about the environment in both developing and industrial countries along with perceptions that the quality of the environment has declined and will continue to decline (Bloom, 1995). Kempton, et al. (1995), cited in Carter (1996), found that environmental values of Americans had a diverse base. They discovered/noted that environmental values have become interconnected with other American values, including religion and parental responsibility to the next generation, and that environmental concern is more universal than previously thought. Of the three institutions most likely to be involved in environmental issues, the public have the greatest trust in scientists, followed by government; they trust industry least. But while they trust scientists the most, they very often get their information related to environmental issues from the media, rather than accepted scientific sources.

Evidence suggests widespread public concern for issues concerning local communities as well as the global environment. The imperative is to provide locally relevant and scientifically credible information through effective educational programs that can provide more depth than the media does. This is exactly the premise under which the education and outreach program for the *National Assessment of the Consequences of Climate Change for the United States* is based.

THE NATIONAL ASSESSMENT

One of the most exciting recent developments in global change research is the launching of the *U.S. National Assessment of the Potential Consequences of Climate Variability and Change.* The 1990 Global Change Act mandates periodic assessment of the consequences of climate change and climate variability for the people, environment, and economy of the United States. The first of these U.S. assessments began in 1997 with a focus on the local, regional, and national implications of climate variability and change. There are three foci of the assessment: regions, sectors, and synthesis. The cross-cutting sectors are forests, water, coastal areas, agriculture, and human health. The synthesis, or overview, document summarizes the information gathered from both sectors and regions. For the regional component, the U.S. is roughly divided into 20 regions that hold some common ecological boundaries; with the 20th region being a virtual organization of tribal lands across the country. The National Assessment is designed to create a continuing dialog among government, business and industry, labor, non-profit organizations, the scientific research and education communities, and the public. Educational materials and outreach programs for the National Assessment will focus on regional issues and impacts with graphic illustrations, mitigating/coping options, and age-appropriate learning experiences. NASA and EPA are providing support for the educational materials development and training and dissemination activities for formal and informal educators. Pending available funds, a full-fledged national educational program will be underway by the year 2000.

INFORMED DECISION MAKING

The educational relevance of this assessment is that it begins to address exactly what people want to know, i.e., what does this mean for me? The UNESCO work plan states that the "... central goals of education must include helping students learn how to identify elements of unsustainable development that concern them and how to address them. Students need to learn how to reflect critically on their place in the world and to consider what sustainability means to them and their communities. They need to practice envisioning alternative ways of development and living, evaluating alternative visions, learning how to negotiate and justify choices between visions, and making plans for achieving desired ones, and participating in community life to bring such visions into effect. These are the skills and abilities that underlie good citizenship, and make education for sustainability part of a process of building an informed, concerned and active populace. In this way, education for sustainability contributes to education for democracy and peace." (1997, UNESCO, p. 24).

Similarly, while it is important to understand how all things are connected on a global scale, only those things that are under one's immediate control will actually be changed. Sometimes, one's immediate control includes far reaching policy decisions for a corporation or a state or a country. For the vast majority of us, however, things under our immediate control concern our daily lives, our homes, our communities, and our personal behavior. This is significant even though it may seem small, because it is individual behavior in the collective that ultimately affects big systems like the atmosphere, or the rivers, or the ocean.

Research on global change education programs concluded that "... providing specific environmental issue knowledge can lead to personal insight with environmental responsible behavior changes as an outcome." (1998, Carter)

Indigenous cultures often carry on the Iroquois tradition of considering seven generations ahead when making decisions. The consequences of decisions made today are discussed in terms of the potential impact they will have 100+ years in the future. Again, thinking in such long time scales is difficult. Leading Native American scholar and author Vine Deloria, Jr. reportedly suggests that Native practices of considering the consequences for seven generations when making decisions may be conceptualized more easily if we see ourselves in the present, in position 4 with three generations behind us (parents, grandparents, and great-grandparents) and three generations ahead (children, grandchildren, and great-grandchildren). We can sometimes remember and usually have photographs of three generations before us and can visualize within our lifetimes seeing three generations ahead of us. This personalizes the decision-making in the same way Native people use family names like Mother Earth and Father Sky, so as to be reminded that we are all members of one family, not just human families but "related" to the Earth.

We are then in a position to consider practices in the recent past that may be useful to reinstitute as we look ahead to potential impacts on the future. The Herman Miller, Inc. furniture company in Michigan, illustrates an example of implementing this approach. They researched past practices and discovered that people crossing the country in wagon trains wrapped their furniture and goods in blankets. The Herman Miller company reinstituted the practice of blanket wrapping for shipping furniture, thereby reducing waste since blankets are reusable, and saving space used by plastic and cardboard boxes allowing them to pack more on the truck.

CONCLUDING COMMENT

Each one of us is important and we can make a difference. Most people experience an innate striving towards the preservation of life and the seeking of quality in life for all living things. To the extent that people spend time in nature, their interest in protecting it increases. Nature has a way of teaching us how to tap into our essential self, to find serenity and beauty, to connect with things bigger than ourselves, to find meaning and solace in the discovery that we are not alone. Who has not experienced this feeling when standing at the edge of the ocean, or a cliff, or a mountain, or a stream, or a vast expanse of prairie, or desert? The challenge is to find this same connectedness in our backyard, and neighborhood, and school grounds, and towns. Understanding how all things are connected, and how one action affects many aspects of an intricate web, can contribute to our ability to make informed choices that help to ensure the quality of life we all strive to enjoy and to enable future generations to do the same.

RESOURCES

Global Change Education Web Sites

U.S. Department of Agriculture

www.fourhcouncil.edu/wenvtop.htm
General agriculture education information on the 4-H site

www.fs.fed.us/outdoors/nrce/welcome.htm
Forest Service Conservation Education site with a link to "Investigating your environment—an activity book for grades 6–12."

www.usda.gov/oce/waob/jawf/
Assessments on the effects of climate on agriculture, maps.

U.S. Department of Energy

www.sandia.gov/ESTEEM/
Education in Science, technology, Energy, Engineering, and Math.

U.S. Department of Interior

www.usgs.gov/education/
The Learning Web at the US Geological Survey.

http://edcwww.cr.usgs.gov/eros-home.html
Access to EROS data laboratory with archive and current images from Landsat satellite.

http://marine.usgs.gov/fact-sheets/indextoo.html
Fact Sheets: Briefings on Coastal and Marine Geology Projects.

www.blm.gov/education/index.html
EE homepage features Sonoran Ecosystems, Dinosaurs, Columbia River, riparian areas, Oregon Trail Interpretive Center.

www.nps.gov/interp/learn.htm
National Park Service "The Learning Place", ParkNet.

www.cr.nps.gov/toolsfor.htm
Links to the Past, Tools for Teaching Archeology, History, Maritime, Museums

www.fws.gov/
US Fish and Wildlife Service Home Page

The U.S. Environmental Protection Agency
www.epa.gov/globalwarming/
Fact sheets on impacts, graphic slide presentation, overview of climate change, national and international policy.

www.epa.gov/epahome/students.htm
Environmental Education site and Explorers Club ages 5–12.

www.epa.gov/enviroed/ftfee.html
Links to environmental education sites in federal agencies.

Global Change Research Information Office
www.gcrio.org/edu/highered.html
Global Change college and university courses on-line.

www.gcrio.org/edu/EarthEducation.html
Links to government and other resources on global change education.

www.gcrio.org/edu/elementary/itselem.html
Educational activities for youth.

National Aeronautics and Space Administration
http://education.nasa.gov/
NASA Education Home Page

www.earth.nasa.gov/gallery/index.html
Earth Science Image Gallery

http://observe.ivv.nasa.gov/nasa/core.shtml
NASA Observatorium with Earth and space data; pictures of Earth, planets and stars; resources; games.

http://gcmd.nasa.gov/pointers/edu.html
Global Change Master Directory, www servers with educational materials about Earth Science and Global Change.

http://spacelink.nasa.gov/.index.html
Activities and Resources for students and teachers.

www.giss.nasa.gov/
NASA Goddard Institute for Space Studies, global warming debate, datasets, and images.

http://southport.jpl.nasa.gov/companion/
Jet Propulsion Laboratory site, new CD "Visit to an Ocean Planet"

http://quest.arc.nasa.gov/index.html
NASA K-12 internet initiative

www.cotf.edu/
Classroom of the Future on-line courses on global change, U. West Virginia.

National Oceanic and Atmospheric Administration

www.photolib.noaa.gov/1b images.photgen.htm
Access to NOAA photo library, with photographs of scenic landscapes, severe storms, coastal areas, in categories but no captions.

www.photolib.noaa/1b images/animals/anim0421.htm
Alaska seabirds

www.photolib.noaa.gov/1b images.animals
All kinds of Alaska animals

www.photolib.noaa.gov/1b images/scenes?coastlines.coast.html
California coast

www.photolib.noaa.bov/1b images/nurp/
Hawaii images

www.photolib.noaa.gov/1b)images/corps/corp2285.htm
Pacific Northwest (also numbers 2284 and 2281)

www.photolib.noaa.gov/1b images/nssl/index.html
Southwest climate extremes

www.ncdc.noaa.gov
Explains decadal and centennial climate change, causes, and whether we can predict.

www.photolob.noaa.gov/albums.html
Pictures of severe storms

www.ncdc.noaa.gov/ol/climate/globalwarming.html
Overview text, what is greenhouse effect, E1 Nino, variability, sea level.

www.elnino.noaa.gov/lanina.html
Animations, explains causes and effects of E1 Nino and La Nina.

www.elnino.noaa.gov/edu.html
On-line workshop

www.oar.noaa.gov/k12pdfs/
Educational activities

www.al.noaa.gov/WWWHD/pubdocs/EnvironIssues.html
Research on the major environmental issues that concern ozone, global warming, El Nino. Includes graphics, IPCC executive summary, FAQ's.

http://globe.fsl.noaa.gov
Site for the GLOBE project, student investigations including El Nino, weather impact on biological cycles, global warming, and desertification.

www.ogp.noaa.gov
Reports to the Nation on-line publications on Climate System, El Nino, and Ozone. Click on ENSO page—graphics. Click education and outreach projects.

www.yoto98.noaa.gov/
Year of the Ocean Web Site, good graphics, images.

www.yoto98.noaa.gov/kids.htm
Learning activities for students, teacher resources.

http://nimbo.wrh.noaa.gov/Portland/edukids.html
A weather page for kids.

National Science Foundation
http://www.nsf.gov/home/her.start.htm
Grants and awards.

http://www.sln.org
Science Learning Network, schools using telecomputing to support inquiry-based science education.

Smithsonian Institution
http://www.nasm.si.edu/earthtoday/
Web site for the Air and Space Museum's remote sensing exhibit focusing on the whole Earth system.

http:/www.si.edu/resource/faq/nmnh/ecology.htm
Assorted environmental and global change sites at the SI.

http://educate.si.edu/educate.html
Teaching resources, professional development for teachers.

Other
www.whitehouse.gov/WH/kids/html/home.html
www.gxnit.com/grabhorn/climatechange/
Graphic images from OSTP publication "Climate Change: State of Knowledge"

http://server2.greatlakes.k12.mi.us/
Explore—collection of fact sheets, articles and classroom activities on global change.

www.col-ed.org/smcnws/msres/
Math and Science Resources, Curriculum Clearinghouses, Lesson Plans.

www.windows.umich.edu/
Windows to the Universe, Earth and Space Science.

www.earthforce.org/green/
GREEN: Global Rivers Environmental Education Network, on-line student investigations.

www.mbnet.mb.ca/lucas/gcg/
The Global Change Game, Environment and International Development Education Program.

www.enc.org/
Eisenhower National Clearinghouse for K-12 math and science resources.

REFERENCES

Bloom, D.E. 1995. *International Public Opinion on the Environment*, Science 269:354–360.
Carter, L.M. Dec. 1998. *Global Environmental Change: Modifying Human Contributions through Education.* Journal of Science Education and Technology. New York and London: Plenum Press; and unpublished dissertation, 1996.
Declaration of Thessaloniki, Dec. 1997. International Conference on Environment and Society: Education and Public Awareness for Sustainability, UNESCO and the Government of Greece.
Department of State. 1997. *Climate Action Report*, Submission of the United States of America Under the United National Framework Convention on Climate Change.
Dunlap, R.E., G.H. Gallup, Jr., and A.M. Gallup. 1992. *Health of the Planet: A George H. Gallup Memorial Survey*. Princeton, New Jersey: Gallup International Institute. Federal

Coordinating Council for Science, Engineering, and Technology (FCCSET) Committee on Education and Training (CET), September 17, 1993. *Report of the Ad Hoc Working Group on Environmental Education and Training.*

Harris, L., V.L. Tarrance, and C.C. Lake. 1989. *The Rising Tide: Public Opinion, Policy, and Politics. Americans for the Environment*, Sierra Club, National Wildlife Federation. Washington, DC.

Hart, Peter D. Research Associates, Inc. December 1994. *Key Findings from a Post-Election Voter Survey Conducted for the National Wildlife Federation.*

Intergovernmental Panel on Climate Change (IPCC), 1995. *Climate change 1995—The Science of Climate Change, Contribution of Working Group I to the Second Assessment Report of the Intergovernmental Panel on Climate Change*, eds. Houghton, J.J., L.G. Meiro Hilho, B.A. Callander, N. Harris, A. Kattenberg, and K. Maskell, Cambridge University Press.

IUCN World Conservation Union Commission on Education and Communication, 1997. *Educating for Sustainable Living: Imagine Tomorrow's World.* International Conference on Environment and Society: Education and Public Awareness for Sustainability, Thessaloniki, Greece.

Kempton, W. 1991. Lay Perspectives on Global Climate Change. Global Environmental Change, June: 193–208.

National Geographic Association, 1994. *Geography for Life, National Geography Standards.* Washington, DC: National Geographic Research and Exploration, ISBN #0-7922-2775-1.

National Research Council, 1983a. *Toward an International Geosphere-Bioshpere Program.* National Academy Press, 81 p.

National Research Council, 1994. *National Science Education Standards*, Washington, DC: National Academy Press.

Report Assessing Environmental Education in the United States and the Implementation of the National Environmental Education Act of 1990, December 1996. U.S. Environmental Protection Agency Environmental Education Division, Washington, DC.

Roper Starch Worldwide Inc. December 1994. *Survey on Environmental Attitudes and Behaviors of American Youth with an Emphasis on Youth from Disadvantaged Areas.*

Washington, DC: The National Environmental Education and Training Foundation. UNESCO, Nov., 1997. *Educating for a Sustainable Future: A Transdisciplinary Vision for Concerted Action*, EPD-97/CONF.401/CLD.1.

United Nations Conference of the Parties, Dec., 1997, *Kyoto Protocol to the United Nations Framework Convention on Climate Change.*

U.S. Global Change Research Program, 1995. *Our Changing Planet*, Washington, DC.

CHAPTER 3

From Policy to Practice
Creating Education for a Sustainable Future

Jack Byrne

In our children we see that education is to the mind what dreams are to the heart. The innocence of children gives them capacity for education and dreams.

—Theodore Strong, Columbia River Inter-Tribal Fish Commission

Today there is a wealth of new thinking about schooling, yet it is fashionable in America to say that schools are failing and there is a groundswell of anger against educators of all kinds . . . It is a mistake to reform the educational system without revising our sense of ourselves as learning beings, following a path from birth to death that is longer and more unpredictable than ever before . . . The avalanche of changes taking place around the world, the changes we should be facing at home, all come as reminders that of all the skills learned in school the most important is the skill to learn over a lifetime those things that no one, including the teachers, yet understands.

. . . The world we live in is the one we are able to perceive; it becomes gradually more intelligible and more accessible with the building up of coherent mental models. Learning to know a community or a landscape is a homecoming. Creating a vision of that community or landscape is homemaking.

—Mary Catherine Bateson, Peripheral Visions: Learning Along the Way

Jack Byrne, Program Director, Center for a Sustainable Future, Shelburne, VT.

Imagine a child going to bed at night—one whose last sensation before dozing off is of deep satisfaction because that day at school he or she made a describable and meaningful contribution to the well-being of his or her community. Imagine that child also feeling excited about going to school tomorrow because learning is fun and school is where new knowledge and skills are gained and shared with others. Imagine a child wanting to do that because it will create a more sustainable future for the generations beyond.

There *are* children who feel this way because they have such experiences. Whenever I meet one I am grateful for them, their parents and their teachers. But we need to assure that this is a more common occurrence based on real experiences and not just an abstraction that this is the way it ought to be. We need to help kids restore and reinforce personal value in learning and acting on behalf of a healthy, prosperous, and equitable future.

Studies show that many of today's youth have little hope about the future. A recent Gallup Poll reports that 70% of 16–24 year-olds believe that the world was a better place when their parents were their age, and 56% said it will be worse for their own children. At-risk youth say they "live for today" and see no hope for their future.

Even for kids with good family situations, the news about the health of the earth's ecosystems and of humanity itself can be starkly negative. We often hear facts from authoritative sources like: "The world's tropical forests are being destroyed at the rate of 100 acres every minute." (editors of The Ecologist, 1987). Or projections like: "If present rates of destruction continue, the tropical rainforest will be completely consumed in another fifty years." There *are* plenty of alarming concerns: depletion of ocean fisheries; habitat destruction; warnings not to eat fish because of toxics and metals accumulation; mysterious mutations of frogs and toads; reports of senseless murders, kids shooting kids, abandoned newborns, and on and on. Such news and images shape our children's perceptions of the world they live in.

The world is full of equally wonderful and miraculous news, but it tends not to make the headlines. Without a foundation of experiences that foster critical thinking and belief that a person can make a positive difference, pessimism and despair are wont to creep in to one's worldview. For roughly one-third of the waking hours in a week that kids are in school, we ought to be doing our best to provide a foundation from which kids can constructively explore, interpret and critique the events in their lives. By doing so, we increase the likelihood that all children will develop a sense of responsibility for the future and the skills and confidence to achieve a prosperous, equitable and sustainable one. As Stapp, et al. write:

> *Schools are a reflection of the society and communities that support them. In a world in which ecological and social problems seem to be rapidly increasing in scope and complexity, our educational institutions should turn toward approaches that generate solutions to these problems.*
>
> *Schools can provide opportunities for the development of skills such as critical thinking, problem solving and collaboratory learning so students can become active citizens with the knowledge to improve their social and natural environments. Students can develop a positive self-concept and become empowered to act by tackling controversial issues that they identify and recognize to be important. In this way, students may begin to feel they are capable and have the ability to improve society.* (Stapp, et al., 1996)

BUILDING A FOUNDATION FOR SUSTAINABILITY EDUCATION

A new project, Education for a Sustainable Future (ESF), is underway in response to the need to engender students throughout the Nation with the skills, vision, and knowledge to contribute to a sustainable, information-rich future. ESF is a comprehensive, nationwide effort to move beyond recently developed policy on sustainable development education to its implementation in the form of good practices at the K-12 grade levels. One of the key hypotheses behind this project is that "learning about the future is fascinating and motivating to all learners and that interest in the future can start with immediate concerns for jobs and vocational skills, move on to central issues of culture, history, and science, and end by engaging everyone in the community in issues of sustainability and personal growth." (ESF proposal, 1996).

Education for a Sustainable Future is a collaborative effort of Concord Consortium's Center for a Sustainable Future, Cobb and Fulton County Public Schools (Georgia) and a number of business partners including Lockheed Martin, IBM, AT&T and BellSouth. It is a five-year project funded by a 6-million dollar grant from the Department of Education Technology Innovation Challenge program and a significant amount of human and technological resources from business partners and others. The project began in October, 1997. Independent evaluation of the project over its lifetime is being conducted by Karen Cohen and Associates.

This project evolves from a considerable body of previous work to clarify sustainable development and its implications for education efforts. Agenda 21, developed at the 1992 Earth Summit in Rio de Janeiro, declared that "To be effective, sustainable development education should deal with the dynamics of the physical, biological, social, economic, and spiritual environment. Information regarding all of these aspects should be integrated into all disciplines" (Sitarz, ed., 1993). In 1993, President Clinton established the President's Council on Sustainable Development (PCSD) to help fulfill the US commitment to Agenda 21. Over nearly three years, the 400 members of the Council developed frameworks and national policy. Consideration of the educational implications of Agenda 21 began in 1994 with the creation of the National Forum on Partnerships Supporting Education About the Environment which was charged to create a blueprint outlining goals and principles of education for sustainability. The basic idea of sustainable development education in the resulting PCSD report is that students at all grades must gain "a basic understanding of the interrelationships among environmental, economic and social equity issues" (Education for Sustainability: an agenda for action). This project continues the effort begun in Rio by developing the first comprehensive US implementation of sustainable development education. (from ESF proposal)

Inquiry-based sustainable development materials in five topic areas are being developed and piloted in nine Georgia schools (K-12). The five topic areas are: Stewardship of Resources, Designing Sustainable Communities, Thinking About and Affecting the Future, Economics, and Global Issues. After careful, independent evaluation, the project will be expanded across eight districts in the South and then disseminated nationwide. The result will be the wide availability of excellent, technologically based materials for learning through inquiry about sustainable development.

Sustainable development is a complex, interdisciplinary topic. Technology-based tools that assist planning, modeling, communicating, collaborating, and decision-making are being created or modified for learning sustainable development, for participating in sustainable development activities, and for entering related vocations. Technology is also providing the collaboration tools to bring international expertise to the schools, to support teachers as they develop and implement the materials, and to disseminate the resulting materials. Based on an educational strategy of student inquiry, we are using a mix of existing general purpose software tools and four new tools developed specifically to help students visualize and explore possible futures.

The objectives of the project are:

1. Define sustainable development education (SDE). Using self-selected teachers, world renowned content experts, and expert staff, we are developing guidelines for integrating sustainable development education across the curriculum.
2. Develop new, technology-rich SDE strategies. Teams of teachers, staff, and consultants will use the guidelines to select and

develop materials that use technological tools to support student inquiry in the five topic areas.

3. Assemble and develop supporting software tools. Eight kinds of computer-based tools are being supported in this project for use across the sustainable development materials. Four of these tools have been selected from the best available software; four that address unique SDE needs are being developed.
4. Implement, evaluate, and revise initial sustainable development education materials. The materials developed by the project are being piloted by the authoring teachers in nine schools serving over four thousand students. The materials will be revised as a result of careful evaluation of these trials.
5. Develop effective professional development strategies. The project is developing a combination of workshops, netcourses, and on-line communities that can reach large numbers of teachers at different grades with background information about sustainable development education, inquiry-based teaching strategies, and ongoing professional support.
6. Disseminate sustainable development education materials to seven collaborating districts. The pilot program will be expanded to the balance of Cobb and Fulton County schools as well as five other districts that will have online access to all the project software and materials. Four hundred teachers in our consortium will participate in a two-year professional development program and netcourses on sustainable development education.
7. Disseminate the materials and approach nationwide. ESF is making all the project-generated material, software, and netcourses available online on a fee-recovery basis that will permit the project to continue post-funding. We will widely disseminate online project information, reports, and evaluations.

ESF will generate educational materials and activities that fully exploit the educational potential of technologies to serve our educational goals. In devising these, we are guided by the following educational principles that are central to educational reform:

Active learning. Students learn best through activities that are engaging and require inquiry and problem-solving. The technological tools being developed will give students unprecedented ability to learn through their own investigations.

Meaningful contexts for diverse populations. General concepts are best learned in a variety of contexts that illustrate the concepts. Because the technology is portable, it will help students learn in any context. Because the approach requires communication with a wide variety of community members, diverse experiences and voices will be heard. The process used will enable students to help define their own meaningful contexts for learning (Center for Children and Technology, 1992).

Depth of learning. It is important to give students time and resources to inquire deeply into a subject so they become expert and understand that they can acquire similar expertise in the future. Networking and computer tools allow students access to far deeper content than previously.

Collaboration. Working in real and cybernetic groups is a powerful learning strategy that also teaches interpersonal skills demanded in the workplace. We are making extensive use of online collaborations and supporting it with the best available groupware.

Communication. Learning to communicate and teach ideas clearly through a variety of media is essential for success and an excellent way to learn. In addition, the technology supports frequent access to mentors and collaborators.

Parent/Caregiver engagement. Involving adults at home in student education is an important way to improve learning and, at the same time, to teach the adults (Epstein, 1995). Portable computers and home networking adds many new opportunities for doing this.

Developmental appropriateness. At the elementary level, proven technology-based products and on-line and off-line learning activities will be used to fortify learning in the core disciplines that are prerequisite to an understanding of sustainable development.

Community links. Students can learn a tremendous amount from younger and older students, parents, grandparents, and other members of the community. Parents, businesses, universities, technical colleges, government agencies, community groups and elderly citizens will meet with students and be connected to them on-line.

As the sustainable development materials are developed and tested, teachers will adopt one of four strategies for its introduction in their curriculum:

Diffusion. Without changing the curriculum, SDE topics will be inserted for illustrations, examples, readings, and assignments. Guided by a framework developed by a study group, we are searching the entire K-12 curriculum for opportunities to diffuse sustainable development material into existing topics and curricula.

Substitution. Units or modules will be substituted for other content in existing courses, addressing objectives of the course in the context of the new material. This is a particularly effective strategy for easily adding new material while minimizing disruption. In each of the five topics, at least two modules of two to six weeks of classroom activities will be developed for elementary, middle, and secondary levels. This totals a minimum of 36 modules across the K-12 curriculum.

Addition. New content that cannot easily be integrated into the curriculum through diffusion or substitution will be addressed in netcourses and open-ended project electives. The project will develop five new elective, one-semester netcourses for students in grades 10–12. Netcourses are courses delivered over the Internet to students at remote sites. They rely on learning in asynchronous collaborative groups guided by assignments and materials provided by network-based software (Tinker and Haavind, 1997). These netcourses will be invaluable to students wishing to gain deeper understanding of sustainable development; the netcourse format makes it possible to fill courses in these topics even if there is insufficient enrollment at any one school. The format also permits us to use online teaching teams that can include experienced classroom teachers, content experts, and technology experts who can participate wherever they happen to be at times that are convenient to them. The netcourses will also be a valuable asset for teacher professional development, community education, and large-scale dissemination.

Extra-Curricular Activities. A variety of extra-curricular activities will be organized to reinforce the sustainable development theme. We will support community service related to sustainable development as part of the students educational experience. All extra-curricular activities that involve interacting with others will be enhanced by the net and student access to networked computers.

DEVELOPING THE MATERIALS AND TECHNOLOGY TOOLS FOR LEARNING AND TEACHING SUSTAINABLE DEVELOPMENT DURING THE FIRST NINE MONTHS OF THE PROJECT

Sustainable development is a complex and challenging concept. Gaining a coherent mental model of sustainable development is essential for teachers to create new materials that are appropriate for a particular grade range and that have some integrity across the grade ranges. We are applying a number of strategies to assure that the final sustainable development units and software tools are outstanding, inquiry based, highly engaging, and that other teachers would want to use or adapt them in their teaching. We are also paying attention to the process employed for getting to these outcomes and noting what works and what does not so that we can provide some insight to others who would like to create their own materials. Strategies include:

- involving world class experts from the sustainable development field who are helping define the knowledge, skills and values that are important to gain from an education about sustainable development and who are providing their expertise to teachers as they develop curriculum;
- a request for participation application process to select for teachers who are highly interested and committed to meeting the challenges of the project;
- a rich and dynamic website that provides a wealth of resources for project staff and teachers and that serves as an electronic "watering hole" in between face-to-face meetings where participants can further evolve the learning community that has been created;
- ongoing involvement and feedback from K-12 teacher participants in the development of the software tools they will use to help teach the curriculum they are developing; and
- providing quality criteria for teachers that will: guide them in meeting high standards for their work, make their materials easily adopted in an already crowded curriculum menu, and assure a solid connection with the state and district curriculum standards.

Expert Involvement and Support

In the project's first months, we brought together a diverse group of people knowledgeable and experienced in sustainable development and education and asked them to define what is important for learners to know and be able to do in relation to the overall concept of sustainable development and in partcicular to the project's five subtopics (Stewardship of Resources, Designing Sustainable Communities, Thinking about and Affecting the Future, Economics, and Global Issues). A summary of these essential learnings is shown in Appendix 1. Those items provide a basic checklist for the final versions of the sustainable development materials and the technology tools that are produced. As teachers apply and refine sustainable development materials with their students, their learning will be fed back to this group and the list will be reviewed and revised accordingly.

Members of this overview group of experts are also involved in the development of more specific course content for the five subtopics. At this time we are developing the materials for the first two of the five topics: Stewardship of Resources and Designing Sustainable Communities. In the sixth month of the project some of the oversight group people worked with teachers, curriculum specialists and project staff for three days to create a course syllabus for each of the two topics to help teacher-developers become more knowledgeable and skilled in relating these subtopics to the bigger topic of sustainability. The work of this group was placed on the project's website and the selected teachers used it to further educate themselves about the topic for which they would be developing materials.

Teacher Recruitment and Selection

Recruitment and selection of teachers for the first round of materials development (for the subtopics of Stewardship of Resources and Designing Sustainable Communities) was by application. There were fifty slots available and between the two school districts over three-hundred teachers applied. The application form included basic information about subject taught, number of years teaching, and internet access at school and home. It also required that teachers apply in teams of at least two and up to four from a particular school. Each teacher was asked to provide a statement of their philosophy of teaching and learning; experiences that qualify them for the project; and examples of projects that they have implemented in their classrooms. The application also required that the principal of the school endorse and sign the application. Project and central office staff at each school district made the selections. As a result of this process we have an outstanding, committed group of teachers. Here's an excerpt from one of the applicants:

> *"I am, after 24 years of teaching, more interested in a 'sane' answer than chasing the 'same' answer to the question of how to best provide an education for a community of learners. I believe that to assist a community as it moves into the future*

to become a part of the global village, we must begin with a community designed and implemented educational experience. The promise of this project lies in the fact that it begins with a restricted local focus. The project's intent is to sustain a community through a future–not just drop it off at the future. Sustaining our own community does not mean isolating it from the world, It means inclusion in the sustained global village. I am energetic and enthusiastic about the problems and possibilities"

Initially, the approach for developing the materials for each of the five subtopics was to be done by a group of experts, teachers, community representatives, and professional curriculum developers. However, we changed that approach to one where the teachers were the primary developers with support from project staff, experts, and community members. There were several reasons for this change. First, we have a highly interested, capable and motivated group of teachers who want to make the project a success. Their ownership of and willingness to constantly improve the materials is most valuable. Secondly, we decided that the materials were much more likely to be embraced and used by other teachers if they were developed by teachers who know the classroom environment best. Teachers could address much more practically the range of student and technology issues that teachers confront daily as they go about their teaching. Finally, we felt that by making teachers the leaders we would increase the likelihood that some of them will become outstanding trainers and advocates for sustainable development education in future years as we disseminate the project's materials and software tools nationally.

ESF Website

The website for the ESF project provides teachers, project staff and other support people with a range of tools and serves as a space where the project's growth and change can be reflected (the address is: csf.concord.org/esf). The website currently has these major features (in varying stages of development):

About ESF—an abstract of the project proposal, biographies of project staff, contact information, and links to related websites.

About Sustainability—resources to help answer the question, what is sustainability?

Resources—a searchable data base of books, videos, websites, curriculum examples, and other resources related to sustainable development education.

Teacher Center—a significant resource for others to access the ESF curriculum material and learn how it was developed by the project. This section of the site includes:

Teacher Team with a brief bio, picture, and email for the participating teachers Professional development information to help teachers and others gain a better understanding of each subtopic and how it relates to the broader topic of sustainability.

Calendar of related ESF events

Discussion and chat section

Special Features—news stories and a showcase of special projects and presentations.

Software—developed by ESF.

Curriculum—units and lessons for K-12 education developed by ESF.

The ESF website is playing a significant role as a communication and development tool for the project. After the teachers were selected we held a one-day orientation meeting for everyone to meet each other, to describe the project goals and timeline, and to learn more about sustainability topics and explore how it relates to the school and community. The ESF website was introduced and teachers learned how to use it. They also participated in the first of a series of chat sessions about the topics of Stewardship of Resources and Designing Sustainable Communities using software available to project teachers from the website.

The biweekly chat sessions were each hosted by a member of the overview committee (mentioned above) or project staff. They were used as a lead-in to a six-day institute held in June where teachers worked with experts to develop the first round of lesson units. The chat sessions had three major benefits: 1) they helped build a stronger sense of unity among the many people involved in the project, 2) they provided a means by which teachers could share ideas for lesson units and how to connect them to a broader community prior to the six-day institute, and 3) they provided a convenient way for teachers to interact with people from other parts of the country with experience and insight that increased their understanding of how the broad topic of sustainability can be translated to specific subjects they teach.

An excerpt from one of the chats is shown in Appendix 2. This was from a session with Bill Mittlefehldt, a high school teacher in Anoka, Minnesota who has been successful in connecting the high school with the Anoka business community in a broadbased effort to create a sustainable Anoka. To make the chat sessions more productive, we emailed a brief description of the session topic and two or three questions that would be help focus the discussion. In addition, some chat leaders also emailed one or two short reading assignments prior to the chat.

Development of Sustainable Development Education Software Tools

Our approach to technology for this project is to support a set of general computer and networking tools that will find multiple uses in the sustainable development materials as well as throughout the K-12 curriculum. The project is supporting eight kinds of tools, relying on the best available software and developing four exciting new packages of particular value to sustainable development education:

1. Network collaboration tools. These are groupware tools for both synchronous and asynchronous collaboration over the Internet that support sharing of any application. They are being used to support collaboration within the project and the topics.
2. Netcourse development and delivery systems. Netcourses rely on learning in collaborative groups, so they use the same collaborative tools described above. In addition, they require a set of administrative support functions for curriculum planning, scheduling, student resources, student evaluation, and grades. This software will be used for professional development and a netcourse developed for each of the six subtopics.
3. Data acquisition and analysis. Portable computers permit a new kind of probeware (Microcomputer Based Labs, developed by Bob Tinker at TERC) that can gather and analyze data in the field. Concord Consortium, currently the leader in this area, is developing (with other funding) intelligent probes, new software and low-cost interfaces for education. The ESF project is using the latest of this software and hardware.
4. Ecological Footprint Tool. Ecological Footprint analysis lets the user visualize his or her ecological footprint (the amount of earth's land acreage it takes to support the user's food, transportation, energy, and housing needs).
5. Decision-support tools. There are a number of software tools used to help groups reach decisions on difficult issues. Techniques include scenario analysis and planning, Delphi techniques, priority-setting, and parliamentary systems.
6. Visual community planning tool. We are creating a new tool that allows students to create 3D visualizations of possible communities. The visual objects in this tool will be linked to data and mathematical models that evaluate indicators about the community, such as the amount of undeveloped land, the average commute time, and the cost of sewage treatment. In

this way, even young students can create attractive communities they can view and then rank according to criteria they develop that relate to the quality of life.

7. Models and modeling tools. Tools for qualitative modeling, such as Model-It, and quantitative modeling, such as Stella, are essential to exploring the long-term consequences of policies and trends. Models that incorporate real-world data and are adaptable to local data are particularly relevant to sustanable development learning. Models for population, land-use, and global climate change are being developed.
8. Online cooperative simulations and gaming. Gaming simulations, where students assume roles and explore how to resolve conflicting responsibilities of these roles, provide powerful educational experiences. In effect, these simulations substitute active learning about an event for a passive, academic exposition of the event. Networking makes many more gaming simulations feasible and can be an invaluable way to make real and explore the consequences of future dilemmas induced by issues like population growth, inequitable development, and resource depletion. We are developing a set of general tools for supporting gaming simulations that involve time-based mathematical constraints (such as limited resources, voting, and financial exchanges). These tools will be used to support several particular simulations, including simulating financial markets, river basin water utilization, and the United Nations.

The ESF project is using the best available software for categories 1–4 and is developing new software for categories 5–8. Each year we are making decisions about the best software to support in each category for the year. We will then fix the software until the next year so the staff and curriculum development efforts will assure teacher and student access to a well-defined set of tools. A summary of ESF project software is included in Appendix 3.

The software developers for this project have a strong desire to create new tools that respond to the needs and creative thinking of teachers participating in the project. The process for doing so was to first provide a summary of the tools and a series of possible student uses and then to ask for teacher and expert response to these ideas. Using those responses the next version of the new software was demonstrated and tested by teachers at a week-long institute in late June ‘98. Their suggestions and ideas based on the “test drive” of the latest versions are being incorporated into the next revisions of the software tools.

Criteria for Curricular Materials Developed by Teachers

We are fortunate in this project to have a highly motivated group of teachers committed to providing their students with the means for meeting the future with knowledge, skills, and values that will ensure the ability to learn the things that no one yet understands. This is the most valuable legacy of an education. To better assure that this comes to pass, we are using an inquiry based approach as a general guide for the materials developed. At the week-long insitute mentioned earlier, teachers were provided with a matrix, or rubric, of criteria which servers as a checklist for assuring that what they develop is consistent the project and pedagogical goals of the school districts. The rubric incorporates the characteristics described above and is also based on a total quality/continuous approach to teaching and learning.

Using this rubric, and lots of interaction with a variety of world class people concerned and experienced in policy issues and community level practices of sustainable development, fifty teachers from Cobb and Fulton County school districts created an impressive set of first draft lesson units at a week-long institute in June, 1998. These units include such creative approaches to sustainable development education as developing a class investigation of the question: What are the ecological, economic, and social effects of a high school football game? This is a wonderful learning opportunity that responds to the need to provide kids with a meaningful context in which to develop and apply new learning about things that matter to them. It also involves the broader community and opens up a diverse array of options for all manner of ability or interest—both on and off the football field. The teachers, who are working in teams, are presently rewriting their materials and submitting them for review to assure they meet the quality criteria of the rubric. During the '98–'99 school year they will be piloting the lesson units in their classrooms, team teaching some of them, and then rewriting them for submission into the ESF Implementation Guide and website for use and adaptation by other teachers.

While it is early on in this ambitious project, we are encouraged by the progress made thus far. As each of the many pieces of the project becomes more fully developed and tested, ESF will be able to provide a rich set of tools and experience to teachers, curriculum developers, community organizations, businesses and others for learning, acting and teaching for a more sustainable future. It is our intent that, as result of these efforts, many more kids and adults will awake each morning with a positive, realistic and hopeful view of the future—and the knowledge and skills to make it happen.

APPENDIX 1. EDUCATION FOR A SUSTAINABLE FUTURE: OVERALL ESSENTIAL LEARNINGS—A BROAD BRUSH

The goal of this project is to help students develop into engaged lifelong learners with the knowledge, skills and values necessary to be active participants in a sustainable globalizing society in the 21st century.

This will demand that every student be capable of enlightened personal action at the individual and community level that takes into account:

- a deep understanding of complex environmental, economic, and social systems;
- recognition of the importance of the interconnectedness of these systems in a sustainable world;
- and, respect for diversity of "points of view" and interpretations of complex issues from cultural, racial, religious, ethnic, regional, and intergenerational perspectives.

To achieve this goal every student will gain the following ***knowledge:***

1. **Knowledge about Systems** (environmental, economic, and social)
   - Systems and the relationship between their subcomponents.
   - Processes by which these systems behave and change. What are the causes of change and what are the effects of change?
   - How these systems change over time.
   - How the elements of these systems are arrayed over the earth and in human society.
2. **Knowledge about Connections**
   - The interconnectedness of present and future political, economic, environmental and social issues.
   - That there are limits to growth on the planet based on a finite amount of natural resources. And that these limits affect the social and economic systems.
   - The role and interconnection of sub components (terrestrial, aquatic, marine, and atmospheric) of our environmental system that support life on earth. This includes the relation of high quality and abundant water, soil, and air essential to support all life.
   - The importance of a great diversity of life (biodiversity) to the long term sustainability of humankind.

- The dependence of humans on our shared natural resource base for a suitable quality life. (Food, shelter, health, aesthetics)
- The relationships of historical communities and societies that foster sustainability, including indigenous peoples and their environment, their economy, and their societal structure.
- The role and interconnection of sub components of our economic and social systems that support the quality of our life on earth.
- The implications of the distribution, acquisition, depletion, and renewal of resources in determining the nature of societies and the rate and character of economic and social development.
- The influences of humans on agricultural, industrial and post-industrial information societies and the impact of each on the natural, social, and economic environment.
- The role of science and technology in the development of societies and the impact of these technologies on the environment, economy, and social structure and equity.
- The philosophies and patterns of economic activity and their effects on the environment, societies, and cultures.
- The implications and trends in the process of urbanization, decline of the nuclear family, end of the cold war, rise in nationalism, borderless information flow, expansion of democracy, water distribution, food production, climate change, population change, and globalizing capitalism and how they impact the opportunity for sustainablity.
- The relationships between energy production and consumption and the implications for the environment, economic systems and social structure in sustainable and non-sustainable societies.
- The relationships between population growth, geographic distribution, age distribution, literacy, and their effects on natural resource use, environmental degradation, wealth distribution, poverty, disease, social harmony and overall quality of life.
- The indicators that can be used to measure the degree to which individuals, communities, and nations are on a path to sustainable development.
- The elements of quality sustainable design in the communities we build and the eco-efficiency of the products that we create.

3. **Knowledge about Multiple Perspectives**
   - The different cultural, ethnic, generational, historical, regional and national perspectives and philosophies concerning the ecological and human environments.
   - The relationships between personal, community, and global visions for a sustainable future.
   - Cooperative international, national, and local efforts to solve global issues, and to implement strategies to achieve a more sustainable future.
   - The implications of the political, economic, and sociocultural changes needed for a more sustainable future for individuals, families, and local and global communities.
   - The processes of planning, policy-making and action for sustainability by governments, businesses, nongovernmental organizations, and the public.

*To achieve the above goals every student needs the following **skills:***

1. **Analysis Skills**
   - To be able to frame appropriate questions to guide relevant study and research.
   - To be able to utilize and create models that define and describe the interactions within and between environmental, social, and economic systems and predict probable interactions and outcomes.
   - To be able to map systems (environmental, economic, and social) behavior focusing on cause and effect and change over time.
   - To be able to create concept maps and timelines describing environmental, economic, and social systems.
   - To be able to observe and identify patterns occurring in the natural (environmental), human built (social), and abstract (economic) realms.
   - To be able to identify data needs and develop appropriate data acquisition protocols.
   - To be able to analyze, interpret, map (including GIS), and summarize common and disparate sets of data.
2. **Communication Skills**
   - Communicate information and viewpoints effectively.
   - To be fluent in the language of systems, from which one can describe environmental, economic and social constructs and processes to others who have different perspectives.

3. **Collaboration and Cooperation Skills**
   - Develop cooperative strategies for appropriate action to change present relationships between ecological preservation and economic development.
   - To be able to develop shared visions with peers, families, and community members about the future.
   - To be able to work as part of a team to solve complex problems.
   - To be engaged as an active member of a learning community, whose aim is to learn for the love of learning and to create a better quality of life for all.
4. **Deep Thinking Skills**
   - Apply broad and diverse definitions of fundamental concepts, such as environment, community, development, and technology, to local, national and global experiences.
   - Develop hypotheses based on balanced information, critical analysis, and careful synthesis, and test them against new information and personal experience and beliefs.
   - To be able to articulate all sides of an issue with a strong basis of understanding the other viewpoints.
   - To exhibit a constant joy of discovery by engaging in creative sleuthing, logical experimentation, problem solving, scientific inquiry, critical thinking, and meaningful reflection.
5. **Decision making Skills**
   - To exhibit leadership in individual or group problem solving to insure that solutions developed are acted upon in a strategic manner.
6. **Use of Appropriate Technology Skills**
   - Be able to use a range of resources and technologies to gather and analyze information, address and define driving questions.
   - Be able to use a wide range of technologies, including email, www, etc. to effectively communicate and coordinate with peer collaborative groups, mentors, teachers, and to members of the larger community.
   - Be able to use appropriate mapping and visualization technologies (like GIS, remote sensing, and similar technologies, inspiration, etc.)
   - Be able to use and apply ESF software tools including the Ecological Footprint Calculator, the What-If Story Builder, and the Visual Community Designer.

7. **Planning Skills**
   - Be able to develop and use scenarios as a tool for strategic planning about sustainable development.
   - Be able to create a "what if analysis" and a systems thinking approach to issues that influence the sustainable systems of the environment, the economy, and social equity.
   - Be able to apply strategic planning and total quality management techniques to successful action taking for sustainability.
8. **Action Taking Skills**
9. **Conflict Management Skills**
   - Be able to work towards negotiated consensus and co-operative resolution of conflict.
10. **Multiple Perspective Skills**
    - Be able to assess the nature of bias and evaluate different points of view.

It is our hope that as more educators become involved in education for sustainability that they will modify and enrich these beginnings.

APPENDIX 2. CHAT EXCERPT. NOTE: BILL IS THE MODERATOR FOR THIS SESSION

[16:55] <bill> Noah is reminding us to head toward the last BIG Question: what types of learning activities would we see if kids were 'properly linked' to the quality of their community or school??

[16:55] <Osborne> kids don't tell their parents because high school kids don't like to admit they have parents

[16:56] <Osborne> Bill . . . MTV

[16:56] <Becky> I wish there was more collaboration between high school and elem

[16:56] <KathyW> Current events activities to keep them aware and see what they want to get involved with

[16:56] <Osborne> yup, becky

[16:56] <Cynthia> I see my math classroom analyzing and graphing water and soil samples, communicating and researching utilizing the internet.

[16:56] <Osborne> hopefully we would start with something as simple as dinner at the table each night

[16:56] <Cynthia> I see collaboration with schools nearby!

[16:57] <Osborne> then we could grow exponentially

[16:57] <bill> That was a great surge of practical ideas. Those activities use skills to build quality into the community

[16:57] <Eva> Yes, Cynthia.

[16:57] <Kent> I think we are much too compartmentalized—social studies, math, english, etc. we need more interdisciplinary units that show all these things being linked.

[16:57] <ecms>. We feel strongly that simulations encourage students to think like scientists, anthropologists, geographers, etc. in a format that they get excited about.

[16:58] <bill> It might help you to build a process of linking with community support people. That could help you with the exponential growth and expand your resource and support base.

[16:58] <Cynthia> I agree- I used math only because that is the subject I primarily teach- my students will tell you I teach primarily project-based and don't really use a textbook

[16:58] <Elaine> We could compare the amount of litter on our school's ground with our feeder middle schools and high schools.

[16:58] <Eva> They love simulations, but even more, they love real situations where they can see actual results.

[16:58] <Osborne> but you need to have the other fields to be good at those simulations

[16:58] <Eva> Yes, Elaine, that's a great activity—it could link all of us together.

[16:59] <Dot> In the lower grades we can begin by reinforcing the three R's reduce, reuse recycle.

[16:59] <Becky> and learning about animal habitats

[16:59] <ecms> or their personal consumption habits

[16:59] <Elaine> kindergarten can weigh the amount of litter we pick up, graph it and hope it becomes less each week.

[16:59] <bill> It seems that your school and city are on the map. Perhaps you could ask the kids to graph or display what problems they wanted to address in class and work to build the resources they need.

[16:59] <Eva> Is the high school structure such that you could do some joint block scheduling and plan with other curriculum areas, Osborne?

[17:00] <Dot> Building on Bill's idea invite community businesses to help students plant trees lost due to . . . many causes (weather, building etc.)

[17:00] <Osborne> sure it can be done Eva

[17:00] <Cynthia> great idea, Bill. Eva, our middle school is block, and yes, I plan with my science teacher, and it would work with other subjects as well

[17:00] <Eva> Bill, that's a great place to start with the kids.

[17:01] <Kent> At Harrison our block is set up English/Math and Social Studies/science. Some things could work there

[17:01] <Osborne> you would have to set aside certain teachers to work with each other for it to be most effective

[17:01] <bill> I think that the sheer volume of INFODATA is overwhelming us. So perhaps our schools need to grow more interdisciplinary to help kids deal with that huge flow (lifetime learners).

[17:01] <Osborne> the density is overwhelming

[17:02] <Becky> Partners in Ed could be a great resource

[17:02] <Eva> The more we integrate, the more connections between "subjects" are seen by the kids.

[17:02] <Osborne> Harrison, ours is set up the same way

[17:02] <bill> Osborne, how close are you to these ed partners of yours?

[17:02] <Cynthia> I think technology is truly the key- If internet were available in the classroom, it would really take us beyond the walls

[17:03] <Osborne> but we run in to the problem that kids say, "well I am not in math this semester"

[17:03] <Eva> Perhaps we need to spend some time with our Partners in Ed this summer.

[17:03] <Osborne> you mean how much do they help us?

[17:03] <bill> Is Ed Partners a local program? Summer is key to getting teachers linked and conceptually focused.

[17:03] <Becky> Many businesses will make a list of resources among their own staff

[17:03] <Cynthia> Or, develop community partnerships related to this (esf) project/plan. Our partners are not active.

[17:03] <Kent> Osborne—we have the same problem. they shut off the subjects they are not taking during the block.

[17:04] <Becky> The partners in Ed program is in Cobb linking schools with businesses

[17:04] <Osborne> we have to get away from "we can't do it that way because we do it this way"

[17:04] <ecms> our partner is very active—providing mentors, landscaping, teacher incentives, and, most importantly, lunch!

[17:04] <Eva> I think we've trained them to "shut off" other subjects. They often still ask, "Does spelling count when I'm writing about something in Math?"

[17:05] <Osborne> maybe we need to rethink scheduling, curriculum flow, etc . . .

[17:05] <bill> In our school, we wrote a book called LINKS which includes 500 persons orgs and internet sources for teachers and kids who want to get beyond the walls (outside the lines). Do you have anything like this>?

[17:05] <Kent> Good point Eva. I find myself guilty of that often.

[17:05] <Sue> Keith again, we have business partners and community partners involved in a big way in this grant . . . this was Sue's comment

[17:05] <Cynthia> Eva, you're right, but persistence can overcome it.

[17:05] <Becky> Bill that sounds great

[17:05] <Sue> They include IBM, ATT, Bell South, Regional Planning, Lockheed

[17:06] <Osborne> what about the smaller companies?

[17:06] <Cynthia> Can we come up with a plan this summer to utilize these partners

[17:06] <Eva> Sue, is there a book like LINKS for listing who can help with what?

[17:06] <Becky> Yes

[17:06] <Osborne> how can they? "help"

[17:06] <Sue> We will be working to provide online mentoring with these business/community partners

[17:06] <Osborne> smaller and more local even

[17:06] <bill> Sue, it might be good to get these Chat partners and others together with the business partners to share visions of the area. Many businesses have grown in their commitment to environmental systems.

[17:07] <Cynthia> I like your idea bill

[17:07] <Sue> This will be an online component by the summer institute

[17:07] <Becky> great

APPENDIX 3. ESF SOFTWARE

Geographic Information Systems (GIS) are used to display, manipulate, and analyze data on computerized maps. ArcView is one popular commercial GIS package. GIS are powerful tools used by scientists, engineers, and policy makers, but can also be very valuable for education. GIS provide many opportunities for data visualization, and for asking questions about relationships and correlations between different types of data. We can help you find satellite images, aerial photos and commercial data sources with information on resource use, demographics, and many other types of datasets. Students can also collect and create their own data.

With GIS, students can:

- Learn about maps—look at a map of the town and find their house, other landmarks. Trace the route from their house to

Existing software

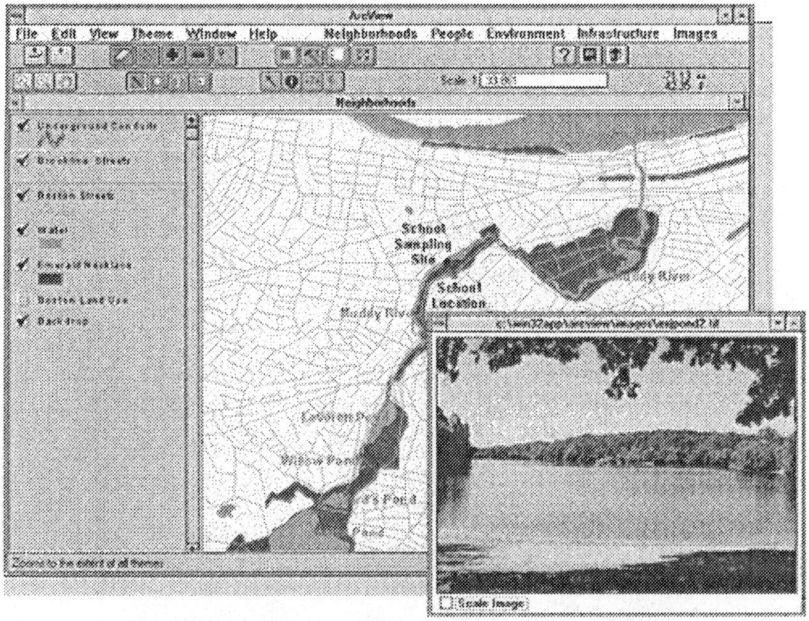

Arc View tool showing ability to link photos with map data.

Figure 3.1. Geographic information systems.

their school. Take photographs of community locations and link them to point locations on a map.

- Learn about how neighborhoods and towns are structured, by analyzing their own neighborhood and using GIS to visualize it in different ways.
- Learn about land uses in their community, calculate total land use by category: residential, commercial, farming, etc.
- Collect data, through surveys, observation, and research within the community, and add it to a map of their community. Locate police stations, fire stations, schools, and think about how these community buildings are located within a community.

Ideas for Sustainability Education

- Visualize energy usage, pollution, population, or other issues over your community. Think about and identify causes, problems, and possible solutions.
- Obtain maps of their community over time, and analyze changes and growth over time.

Existing software

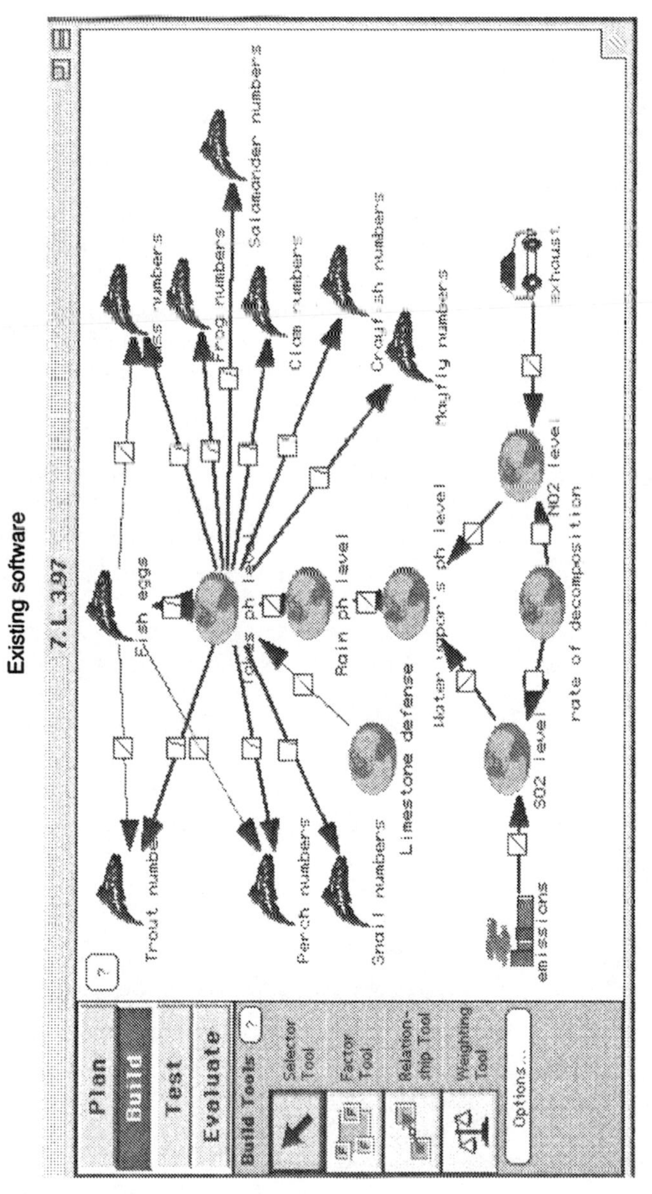

A student's Model-It model of the effects of acid rain on aquatic life

Figure 3.2. Model-It.

- Conduct research on real-world issues, study the relationships between the distribution of urban vegetation, soil types, animal life, air and water characteristics, the built environment, and the human population. (e.g., One group of students in Boston is focusing on using GIS to analyze water quality data along local urban river.)

Modeling is essential to developing an understanding of systems and to exploring the long-term consequences of policies and trends. Running a model built by experts can be a powerful tool for learning about a system, but creating one's own model is equally valuable, in different ways. By building models, students can learn to think about and recognize systems, identifying variables, proposing and exploring cause-and-effect relationships, making predictions and testing their models, presenting their ideas to others, and developing and refining their understanding.

Model-It is a learner-centered tool for building dynamic models. It provides a qualitative, conceptual interface for creating models, so that students can easily learn to use the program, and can focus their attention on expressing and testing their ideas about the systems they are modeling. Model-It lets students express relationships simply, for example, by selecting choices from a menu to fill in the blanks in this sentence: "as the rainfall increases, the stream depth *increases* by *a little*." As students gain expertise, they can define more complex relationships as well. The software has been used with great success by middle and high school students.

Model-It is existing commercial software. Shari Jackson, a member of our project team, is the developer of Model-It, and would be happy to work with curriculum developers to design Model-It activities to integrate with different curriculum areas.

Ideas for Sustainability Education

- Students can build models on any subject they are studying, as a way of representing and exploring their ideas in-depth. Model-It has been used the most for models of stream ecosystems and global climate change, but is equally suitable for any number of relevant sustainability topics.
- Models that incorporate real-world data and are adaptable to local data are particularly important. Modeling should be used in conjunction with hands-on activities, and students can enter the data they have collected into their model.

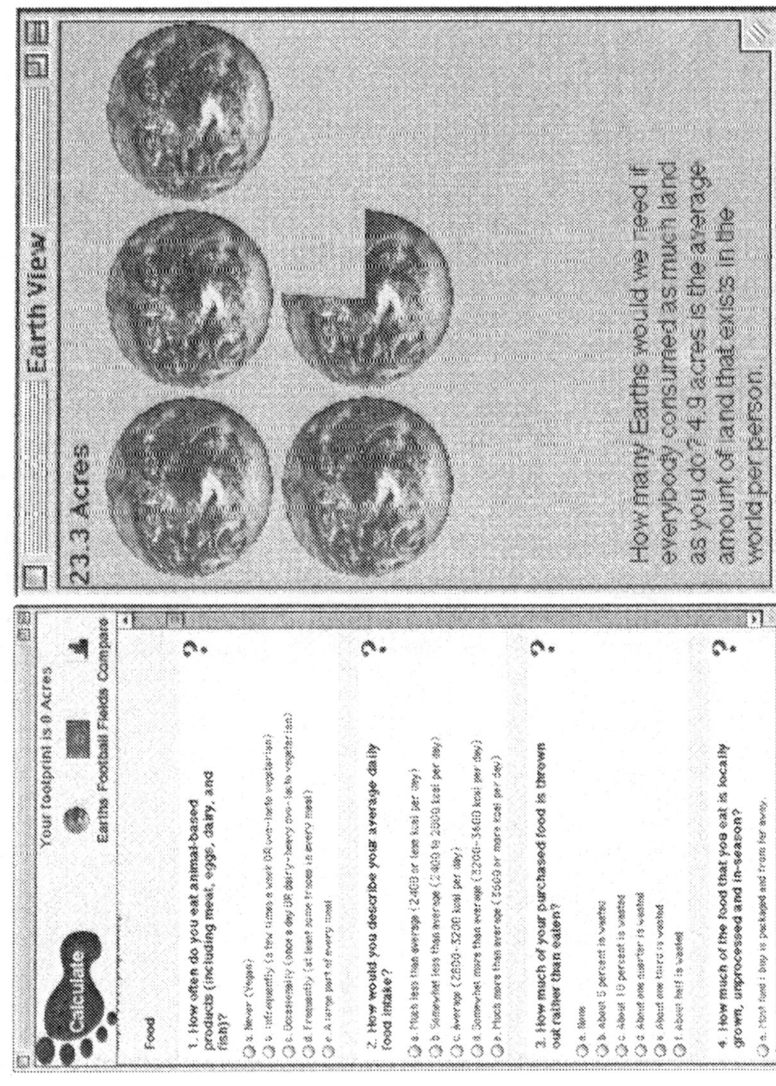

Figure 3.3. Ecological footprint calculator.

The Ecological Footprint Calculator measures our use of nature, by calculating how much land is required to produce all of the resources we consume, and absorb all of the waste we produce. The concept is described in the book, *Our Ecological Footprint: Reducing Human Impact on the Earth*, by William Rees and Mathis Wackernagel.

We are developing the *Ecological Footprint Calculator* as a tool for students to calculate their own ecological footprint, and to visualize their footprint in various ways (e.g., how many earths we would need if everyone on Earth had a footprint like theirs). The tool will also provide information to students about how the ecological footprint is calculated, and what are the different land uses that make up their footprint (crops, grazing land, forest, developed land, etc.). Furthermore, students will be able to use the tool to open up different surveys, to create their own footprint surveys,

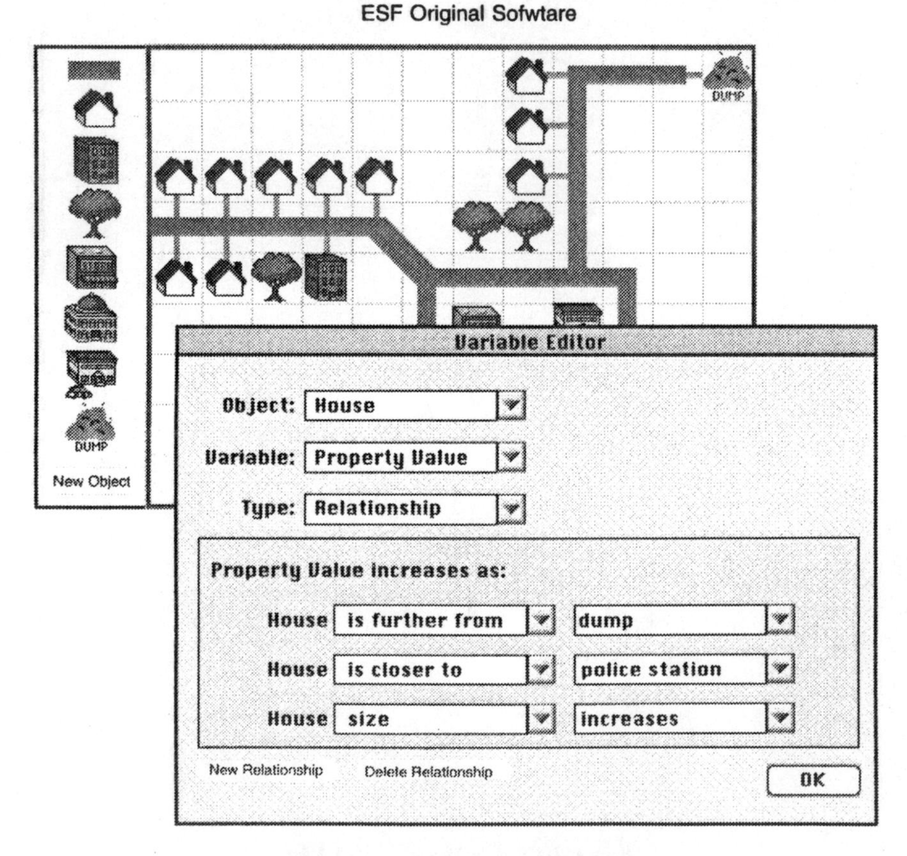

Figure 3.4. Visual community planner.

to edit and create new questions, and to research for themselves the impact of human activities.

We are designing a spatial modeling and visualization tool for community design and evaluation. Each component in the model (e.g., houses) would have variables associated with it. The variables could be viewed and defined by students, and some might be calculated from relationships with other factors, as in the figure above. Students could also do calculations over a community (e.g., total population, electricity use, jobs available, unemployment rate), and perform "what if" experiments (e.g., the net effect of everyone turning their thermostat down a degree, or using public transportation). The tool would provide a variety of ways to visualize, analyze and evaluate a community.

Ideas for Sustainability Education

- Design their own neighborhood and rank it according to criteria they develop that relate to the quality of life, or calculate a "sustainability rating" for it.
- Create a model of their own community, using real data and maps on which to base the model, and evaluate it according to various indicators and metrics. Propose and determine the impact of potential land use changes, like building a new school.
- Calculate and visualize the ecological footprint of a house, school, or community.

This idea is for a general-purpose tool to support and teach the decision-making process. The system would work as follows: A scenario would be presented. Students would be assigned roles. Students would interact and discuss the topic, from the perspective of their roles. First they would decide on goals and options. They would discuss possible outcomes, and their probabilities. Votes would be made, and the result would be entered into a model that would then run the simulation for some period of time, and display the results. The cycle would then begin again.

The system would have two parts: a role-playing interface to support and guide students in the negotiating and decision-making process, and an underlying model with which the students interact. The model could be some existing model, or a model that the students create themselves using software that we develop.

Goals

- Students would explore how to resolve the conflicting responsibilities of their roles.

ESF Original Software

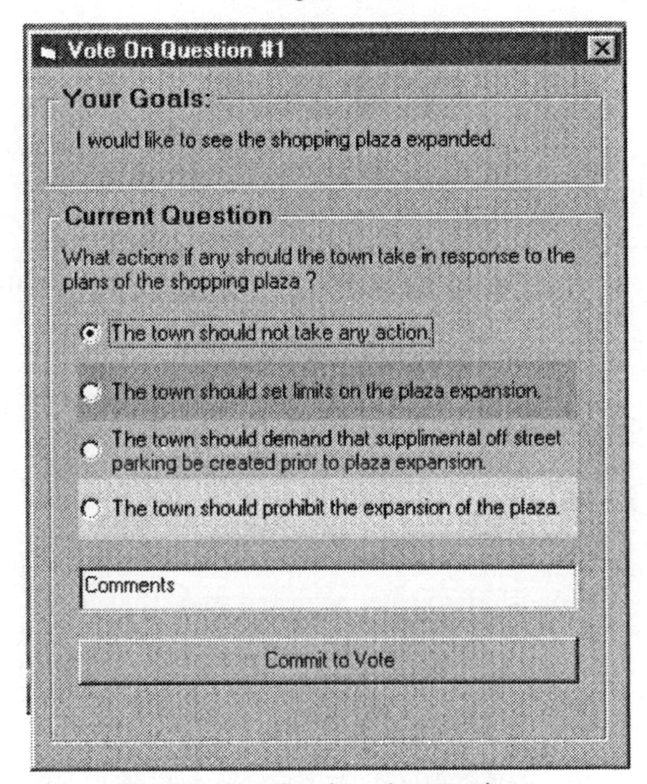

Example of interface for voting on an issue

Figure 3.5. Cooperative simulations and decision making.

- A networked interface would make more simulations feasible and "real".
- Issues such as population growth, inequitable development, and resource depletion could be explored.
- Time-based mathematical constraints would be imposed (such as limited resources, voting, and financial exchanges).
- The system would be general enough to simulate a variety of topics, including financial markets, river basin water utilization, and the United Nations.

This is a tool to create scenario models, one kind of model the Cooperative Simulations and Decision Making software might use, and is based on the Decisions, Decisions software of Tom Snyder, Inc. That software leads students through an interactive, description-based model. The model

ESF Original Software

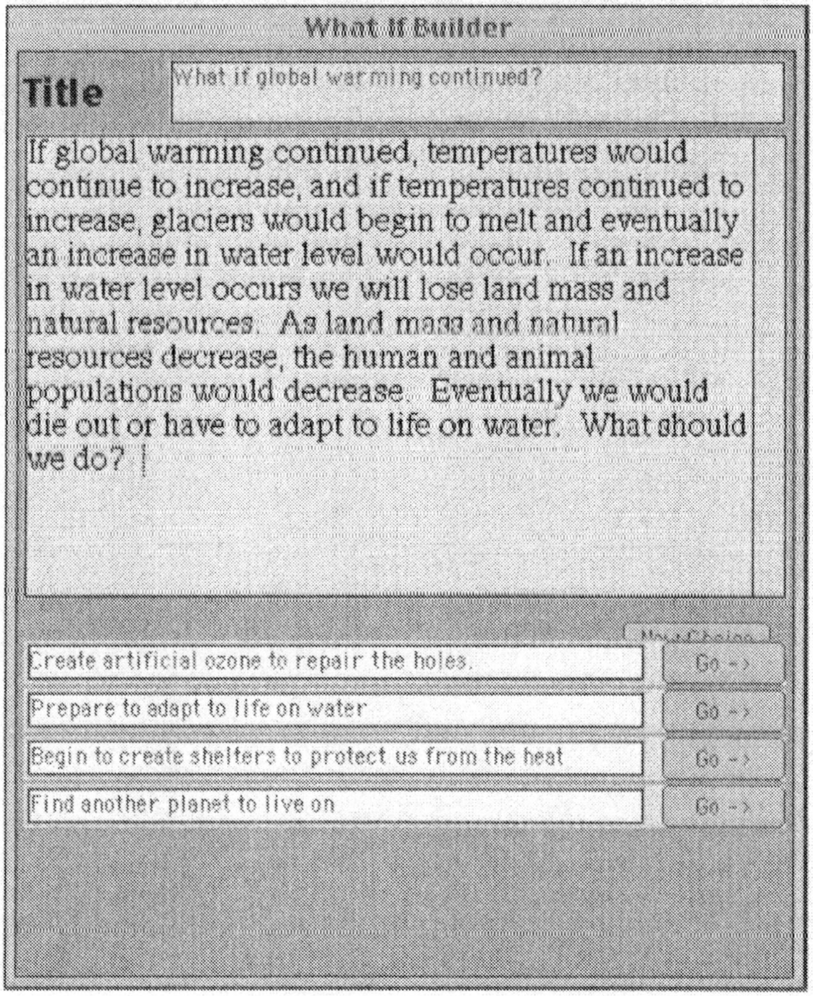

Figure 3.6. What If" scenario builder.

presents a scenario and an issue to be decided, then prompts students to make a decision. Students role-play and debate the issue, and then choose a decision. Decisions, Decisions comes with pre-packaged models for students to explore; our idea is to develop a model-builder for this kind of model. Teachers or students could use the tool to develop their own models exploring the impact of different decisions about a scenario of their choosing. Students could use a model the teacher built as a learning exercise, or

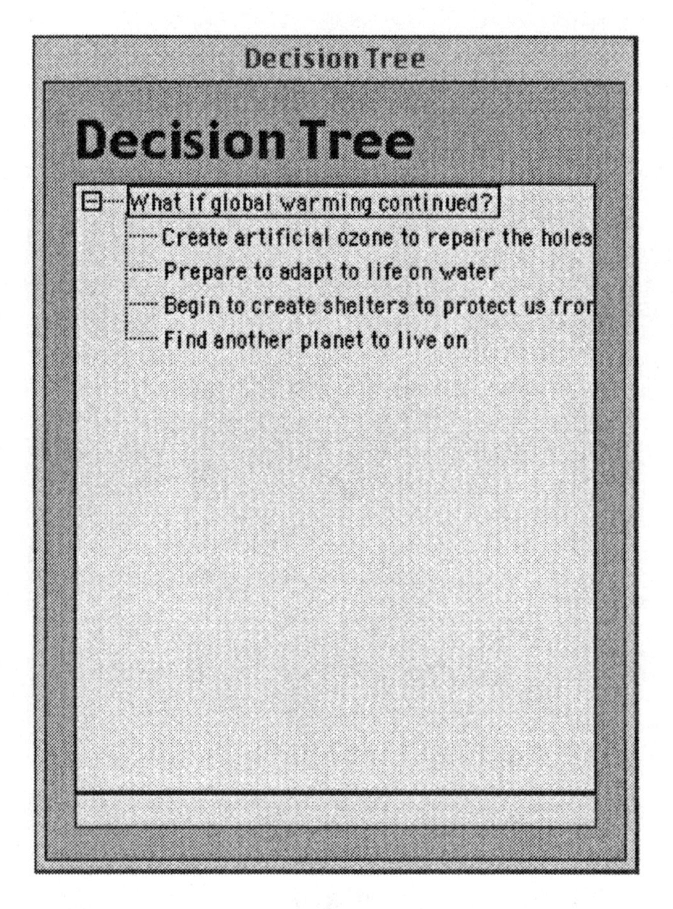

Figure 3.7

create models to share with the class or others. Students could discuss and debate both the decisions to be made, and also meta-issues about the model itself—are these the only possible choices, and is that really a reasonable outcome of that choice.

Ideas for Sustainability Education

- Could be used for just about any topic. Promotes visioning and thinking about the future. Also encourages discussion and debate, and shows that models have built-in assumptions that may be challenged. The game-like concept (create a model for your friends to play) may also encourage students to do more research on investigating the probable impact of different scenarios, in order to create better models.

APPENDIX 4. UNIT DESIGN RUBRIC

CURRICULUM: HIGH EXPECTATIONS			
FOCUS	1	2	3
Lesson Overview	Though the concept, issue or problem is stated, the overview is incomplete and not fully developed.	The overview describes and justifies the concept, issue or problem but lacks coherent description of learning opportunities.	The description of the lesson is clear, concise, focused, and thorough. An overview of learning opportunities in support of concept, issue or problem is included.
Learning Objectives	Learning objectives are unclear or poorly stated; appear disassociated with concept, issue or problem and not significant; seem unrelated to ESF topic standards.	Learning objectives are clear, significant and linked to ESF topic standards.	Learning objectives are compelling and provide the focus to drive student inquiry and interest of the concept, issue, or problem. They are related and supportive of ESF topic standards.
Alignment with standards	The alignment is contrived.	The alignment with QCC and ESF topic standards is clear and explicit but is not embedded in the learning events for students	The alignment with QCC and ESF topic standards is clear and explicit throughout lesson/unit. Learning events are supportive of students' attainment of the QCC objectives and ESF standards.
Classroom Preparation	Minimal information conveyed for preparing classroom.	Preparation notes include materials needed, computer requirements and resources.	Preparation notes indicate helpful information on classroom organization, resources/materials/software/bookmarks and technologies needed.

CURRICULUM: HIGH EXPECTATIONS (*cont.*)			
FOCUS	1	2	3
Learner Preparation	Minimal information conveyed for preparing learners.	Preparation notes include prerequisites for lesson.	Preparation notes include possible misconceptions and ideas of preexposure for learners.
Time Frame	Time allocations are unrealistic or are not clearly indicated for various parts of the unit.	Time frame is difficult to identify or follow for various learning events, instructional activities, and performance tasks, as well as for unit as a whole.	The suggested time frame clearly specifies, in hours, duration of learning events (i.e., introductory activities, instructional activities, performance tasks, assessments) as well as estimated time for entire unit.
Resources/ Bibliography	Resources used are limited in scope and depth and are not related to lesson's focus; information is lacking on technology tools in support of learner inquiry; bibliography insufficient for preparing teacher for unit.	Resources, including technology tools, are varied and directly related to lesson's focus; bibliography includes resources/ bookmarks for content and process information for teachers.	Resources/Bibliography are up-to-date, span a wide range of forms, media, and technologies that directly support students' exploration and inquiry of multiple perspectives related to the lesson's focus; resources support teachers' needs for background information in preparation for unit implementation.

INSTRUCTION: OPPORTUNITY TO LEARN AND TIME-ON-TASK			
FOCUS	1	2	3
Rigor	ESF curriculum unit lacks rigor; focuses students on recall of isolated concepts, skills or facts.	ESF curriculum unit enables students to develop a basic understanding of a concept, problem, and/or skills.	ESF curriculum unit requires that students engage in a thorough exploration of a theme, problem, issue or question. The degree of difficulty and the complexity of content are appropriate for learners.
Relevance	Learning opportunities are presented in ways that prevent students from making meaningful connections between their own experiences and the content.	Learning activities and events enable students to develop an understanding and use of knowledge and skills related to issue or problem. Tasks allow students to derive personal meaning from material.	Learning opportunities require students to engage in a through exploration of topic and draw upon students' interests, backgrounds, cultures and experiences.; learning opportunities include the perspective of time using "7 generations thinking".
Performance Design	The performance task, project or performance assessment is not relevant to ESF curriculam.	The performance task, project or performance assessment is connected and relevant to ESF curricula/topic concepts but lacks one or more of the five essential components listed in column 3.	The performance task, project or performance assessment is (1) feasible, (2) worthwhile, (3) contextualized, (4) meaningful, and (5) sustainable. The performance design is rich enough to help students learn ESF curriculum outcomes.
Degree of Inquiry	The learning activities and events demand no inquiry or research on the part of the student.	The learning opportunities demand some investigation or research on the part of the student but mostly of the nature of finding out facts.	The learning opportunities demand that students search for in-depth understanding through systematic research and inquiry using a variety of sources, research strategies and technology tools.
Use of Technology	Technology is not an integral, integrated component of the ESF unit.	Technologies are used occasionally but inquiry is not dependent on tools; use of technologies is tightly controlled; sequenced and supported tasks leading to independent use of technologies are not evident.	Learning ESF content depends on infusion of technologies through instructional, sequenced learning opportunities in support of student inquiry. Use of technologies is appropriate for the tasks.

INSTRUCTION: OPPORTUNITY TO LEARN AND TIME-ON-TASK (*cont.*)			
FOCUS	1	2	3
Sequence of Content and Process	The unit has no clearly defined structure, or the structure is chaotic.	The lesson or unit has a recognizable structure with substantial content subsumed within the performance design.	Flow of activities is logical and likely to engage students in meaningful activities; concepts are carefully sequenced and integrated with substantial ESF content; learning opportunities support several learning styles/intelligences. Learning opportunities from one part of the unit connect with other parts; unit includes direct teaching lessons as appropriate.; students explore a topic from many different angles and understand the relationship of the parts to the whole.
Benchmark Lessons	Specific teacher designed tasks or activities to develop identified concepts of skills are not well developed in ESF unit.	Teacher designed tasks are included but not sufficient to address anticipated subtropics.	Teacher designed tasks/activities are sequenced and integrated through the unit.
Systems Thinking	Learning opportunities do not include activities to engage students in understanding of the systems that underlay the essential question/issue.	Learning opportunities involve students' in exploring the components of the system and what regulates the system.	Learning opportunities engage students in understanding and describing systems—the components and internal/external controls—underlying the essential question/issue; activities will engage students in exploration of interconnectedness of the environment, economy and the quality of life with understanding how each one affects the others.
Grouping Strategies	Unit addresses either individual, collaborative, ore competitive learning exclusively or forms of learning are not connected in meaningful ways to study of content.	Learning opportunities have students working individually and in groups but learning from these forms is not maximized or linked in ways shows evidence of content mastery.	Learning opportunities include individual, collaborative, and competitive tasks. There is both individual accountability and group interdependency. There are opportunities for students working with students and with teachers as well as students and teachers working with the community.

ASSESSMENT: FREQUENT MONITORING

FOCUS	1	2	3
Timing	Formal assessment is limited to end-of-lesson/unit activities.	The lesson/unit includes diagnostic as well as end-of-unit/lesson assessments	The lesson/unit is assessed from beginning to end in ways that support and measure student learning, inform teaching and inform the learner.
Assessors	Only the teacher evaluates students' work.	Student is asked to reflect in general; students interact with peers to share and give feedback but there is limited use of rubrics and checklists.	Assessments/Artifacts include measures that guide student self-assessment and reflection on both products and processes (example, ongoing specific questions, checklists, rubrics); students may evaluate their own and each other's work.
Alignment	The assessments are unrelated to ESF curriculum; assessments do not measure student learning from the curriculum taught.	Only some aspects of the ESF curriculum are measured.	The assessments/artifacts are derived from curriculum-embedded learning opportunities that measure and support learning.
Evidence of Mastery	The assessment requires minimal response from student limited to answers to multiple choice, true false questions, or yes/no oral responses.	The assessment requires some verbal/written communication on the part of the student. This communication is limited to short test answers or question based oral responses.	The assessment requires an elaborate response of both knowledge/skills gained and process. This communication is provided through written, artistic, oral performances, exhibitions, artifacts and/or opportunities for students to teach.
Audience and Purpose	The teacher is the only audience and the purpose of assessments is to test for a grade.	The purpose of the assessment/development of artifacts is vague or only school related.	The performance task, project, or performance assessment asks students to work for a real audience and purpose so that they can experience the benefits and consequences of their work.
Ongoing Feedback	Feedback is very general or ambiguous and given after assessments are completed.	The assessments include measures that provide specific feedback from the teacher. Revision is allowed but not encouraged.	The assessment s include measures that provide elaborate and specific feedback throughout the process from both the teacher and peers. It includes measures that encourage all students to revise in order to produce quality work.

Figure 3.8

REFERENCES

Bateson, Mary Catherine (1994) Peripheral Visions: Learning Along the Way. New York: Harper Collins Books.

Cobb County School District and Concord Consortium (1996) Education for a Sustainable Future: Proposal to the US Department of Education Technology Innovation Challenge Program.

Fien, J. ed. (1996) Teaching for a Sustainable World. Nairobi, Kenya. United Nations Environment Programme (UNEP).

Dewey, John (1910) How We Think. Toronto, Ontario: Dover Publications.

Epstein, Joyce L. School/Family/Community Partnerships, Caring for the Children We Share. *Phi Delta Kappan*, May, 1995.

Krajcik, J.S., Blumenfeld, P.C., Marx, R.W., and Soloway, E. (1994) A collaborative model for helping teachers learn project-based instruction. *Elementary School Journal*, *94*, 483–497.

Milbraith, Lester W. (1989) Envisioning a Sustainable Society: Learning Our Way Out. Albany, New York. State University of New York Press.

President's Council on Sustainable Development (1996) Education for Sustainability: an agenda for action. Washington DC. US Government Printing Office.

Sitarz, Daniel ed. (1993) Agenda 21: The Earth Summit Strategy to Save Our Planet. Boulder Colorado: Earth Press.

Stapp, William B., Wals, Aren E.J., Stankorb, and Sheri L. (1996) Environmental Education for Empowerment: Action Research and Community Problem Solving. Dubuque, Iowa: Kendall/Hunt Publishing Company.

Wackernagel, Mathis and Rees, William (1996) Our Ecological Footprint; Reducing Human Impact on the Earth. New Society Publishers, Gabriola Island, BC, Canada and Philadelphia, PA.

World Commission on Environment and Development (1987) Our Common Future. New York: Oxford University Press.

CHAPTER 4

Educating for Sustainability in American High Schools

Mark DiMaggio

Readers of this book almost certainly do not need convincing that:

a. the sustainability movement is real and urgent, and
b. high schools in the United States are under real pressure (and close scrutiny) to achieve reform which results in higher standardized test scores and (presumably) increased student learning. This focus on raising test scores typically results in the creation of an atmosphere of competition within and between schools.

In this essay I will suggest that there exists an opportunity to link these two phenomena in a new educational system that will advance the precepts of sustainability, dramatically increase student learning and test scores, and replace the climate of competition with one of cooperation.

The Earth's environment and the business, political, and economic systems of the world are changing at an unprecedented rate. The physical and ecological limits of the Earth are being rapidly approached, and some have been exceeded already. In fact, the reality of the 21st century will be

---

Mark DiMaggio, High School teacher of Earth and Environmental Sciences, Paso Robles High School, Paso Robles, CA.

so entirely different from any other era of human existence that traditional methods of thinking and problem solving will not be adequate to meet the challenges ahead. Focusing our teaching efforts simply on raising test scores surely will not solve the problems that are looming over the horizon.

Unfortunately, the approach currently taken to raise standardized test scores in public high schools is fast and furious and in some sense, artificial. Administrators feel that test scores must be raised "immediately" (as in *this* school year) to satisfy parents and the community that rapid, tangible steps are being taken to either stay on top, or to at least move closer to the top. I use the words "unfortunately" and "artificial" because the higher scores likely to be achieved will not necessarily reflect a deeper understanding of the world in which the graduate is soon to find him or herself, but rather a superficial memorization of facts and details, a result of being "taught to the test." As a result, many educators fear that we are not preparing our young people to assume responsible roles in which they contribute to the establishment of a sustainable society.

Happily, and perhaps obviously, there is a wonderful potential synergy which can be nurtured that would raise test scores, provide real understanding for students, and create greater hope for the creation of a sustainable society.

PORTRAIT OF AN AMERICAN HIGH SCHOOL STUDENT

It is unfair and unwise to place people into arbitrary categories. There are no doubt thousands of possible groupings into which adolescents could be placed; there are students who are under-challenged with high school curriculum, students who need continuous special help to pass their classes, students in school strictly to be on a team or in the school band, and students who would prefer to be *anywhere* but in the classroom. But despite these obvious differences, most experienced classroom teachers would agree that certain characteristics do exist that tend to be common to a majority of students. These characteristics would include:

1. Lack of a meaningful parental support system. There are many, many excellent, dedicated, effective single parents. But there are also many ineffective, uninvolved, incompetent parents—single or otherwise. High school is difficult and challenging—much more difficult than when most adults were in school—and at least in my experience (12 years public high school science teacher) well over 50% of the students either

have parents who don't care or who have given up, or are too busy with job obligations or career advancement to provide the support necessary for their kids.

2. Students are under tremendous pressure. Some feel intense pressure to conform to their peer group who regularly engage in at-risk behavior such as drug and alcohol abuse. Many of these same students are strongly encouraged to take the most difficult courses and to achieve excellent grades. There is pressure to play on a sports team or two, volunteer in the community, be an active member of a school club and participate in other after school activities—all to obtain entrance into the top college. From a young age, students are told repeatedly that with a good education they can earn a high salary (more than their parents) and invariably attain security and happiness. In short, college bound students (read *most* students) are overwhelmed with preparation for college, indeed, from a young person's perspective, there really is no alternative to superlative grades in high school, entrance to a top college, obtain the high paying job, and settle into a career. A recent high school graduate of mine told me that his freshman year at the University was a "breeze compared to high school"!
3. Students are fearful for their future. High school age people understand enough about their world to recognize that the current path humanity is following is destructive to society and the environment and is unsustainable (although most wouldn't know that word to use it). *Many adolescents actually do not expect to live much past 30.*
4. Students are disillusioned with the system. The educational system which requires they do 2–3 hours of homework per night beginning in the 6th grade (did *you* do 2–3 hours of homework when you were in 6th grade?), robs them of their childhood (although at this point in their lives they are anxious to grow up). They are disillusioned with a system that encourages them to earn more and obtain more material goods than their parents, and yet these are the parents who are on a 60–70 hour work-week treadmill which allows them no time to know their children. Is this the solution, they wonder? To earn and obtain more and consume more? Young people sense this conflict—a manifestation of the system they operate within—and their frustration results because they see no alternative for themselves.

PORTRAIT OF THE SYSTEM IN WHICH THE STUDENT OPERATES

The high school system attempting to deal with these realities is ill-equipped to create the level of change necessary to generate real, meaningful, powerful reform. The great trend in academics and therefore in education over the last 300 years, unconsciously adopted by educational institutions worldwide, is reductionism. Curricula are fragmented and isolated into their specific categories, creating ever more refined knowledge. This has produced truly remarkable technological achievement, but with the concomitant loss of an understanding of the big picture—the system of nature within which all other subsystems operate. We are graduating students who understand the fundamentals of economics, but have no notion of the basic laws of ecology or thermodynamics, students that have memorized the order of the planets, but do not understand their relative motion or even that they are not "all lined up" like it shows on the poster!

High schools of course isolate curricula in response to college entrance requirements which shape and drive the high school programs. In addition, they are under pressure to produce students capable of earning high SAT scores. Consequently, high schools are under enormous public pressure to graduate as many students as possible ready to enter a four-year university directly after graduation. Indeed, the public image enjoyed by the high school is directly related to their success in achieving this goal.

It certainly is not an unworthy goal in and of itself to graduate students ready for college, but I fear that in our haste to produce high test scores, we are creating citizens who lack a broad perspective on the directions humanity is moving (although as alluded to earlier, they do have an underlying uneasiness that things can't keep going this way for much longer). Be reminded of what David Orr in his book *Earth in Mind* declares; that "the methodical destruction of the natural world on which our civilization depends is accomplished predominately by educated people." Our educational system is currently producing graduates who are increasingly effective at exploiting the resources of the natural world, but continue to operate under the archaic premise that resources are unlimited.

At the same time, we must remember that American high schools are attempting to restructure their approach to provide more vocational training, create more community and business ties and to make learning more relevant. These are excellent goals, and yet they are difficult to achieve in the absence of a framework or context to provide the relevancy. *I here suggest that no better integrating theme, no better method to bring meaning and relevancy, no better approach to creating community and business ties exists, than that of educating for sustainability.*

SUSTAINABILITY AS AN INTEGRATING THEME

It is well known that far greater learning occurs when knowledge can be transferred across disciplines and when that knowledge is perceived as having high relevancy to the students' life. Yet given the current social and ecological state of the world, it is clear that we need a broader, more forward-thinking, more ecologically grounded, holistic vision to guide our educational efforts in American high schools. We need to recognize that our well intentioned, but short-sighted goal of higher test scores is inadequate, and ill-suited to the reality of our world. We lack a vision that connects our curricula into a broader purpose for our students, our community, and our world. Education without a unifying theme will only serve to divide our students and undermine our efforts to provide an excellent educational experience for our young people, and to actively improve prospects for all people.

I believe we need to rethink our stated task of "getting kids ready for the 21st century." What does that mean? What are we training our young people to do? Are we not training them to continue on the course humanity is presently following? Are we not training them to do the same things we have done, only on a grander scale? The reality of the 21st century will be so entirely different from any other era of human existence, that traditional methods of thinking and problem solving will not be adequate to meet the challenges ahead. A psychologist once said that a definition of insanity is doing the same thing over and over again and expecting a different result. Is more and bigger of our "growth approach" to solving problems going to bring happiness, peace and justice to our world? The reality is that the growth of both the human population and the human economy are in *fundamental conflict* with the ability of the Earth to supply the resources and to absorb the wastes that result. Unless we precisely define our vision of what we want the world to become, and actively teach that to our students, our world will continue on its present course. It will become what we teach our children to make it. Given the reality of the new global economy and the extent to which many of our actions produce ripple effects across the planet, our vision of sustainability should extend to the four corners of the Earth, and should include not just humans, but *all life*. High Schools need a theme, a common purpose to bind their programs together, and to give us all a shared vision towards which we all are working.

I propose that this theme should be *global sustainability*—the concept that each generation of people leaves to their descendants equal or better prospects than they had for a healthy, hopeful, prosperous life. Global sustainability can only be achieved when the citizen participants see the world from a broader, more holistic, systemic perspective.

Therefore, rather than focusing on training our students to perform well on standardized tests and out-compete their peers, we should teach them to cooperate.

Adopting sustainability as an operating premise creates an atmosphere of very high relevancy for students. They immediately see why this is important and recognize the implications for their own future plans. This of course results in a strong interest in the curriculum, with powerful learning as an end product. I shouldn't use the phrase "end product" because these students are now motivated to be active participants in the creation of a sustainable society because they understand how it will improve the quality of their own lives.

Consequently, for these reasons, I believe the operative words that guide the restructuring of our high schools should be trust, respect, family, responsibility, and caring (as described in the excellent high school program *Community of Caring*), to which we should add giving (as opposed to taking), love, honesty, cooperation, compassion, and systemic, holistic thinking. Continued competition on a larger scale, will only serve to magnify the already overwhelming social and environmental issues we now face.

A 10 YEAR CLASSROOM EXPERIMENT

Over the last 12 years I have had the good fortune to teach Earth and Environmental Science at Paso Robles High School in California. One of the best things about my assignment has been that I have had some freedom to select the units of instruction I share with my students.

Over my first two years teaching I was struggling to simply stay a day or two ahead of the class, and I stuck with earth science topics that I was familiar with. But as I gained experience, I began to integrate more and more environmental science topics into the curriculum, and was overwhelmed with enthusiasm from the students. Their common response was "why hasn't anyone ever taught us this before?" and "this is the most meaningful course I have ever taken." As the years rolled by and I gained more experience, my course has essentially evolved into a year-long class on Global Sustainability—this because I realized that the theme of sustainability provided the framework students needed to link the various topics we study. These topics range from economic globalization to astronomy, from desertification to social and environmental ethics. Viewed from the perspective of regard for the well-being of future generations, these seemingly unrelated topics suddenly take on more meaning, relevancy, and importance.

I have now developed a course which provides what I consider to be the essentials of ecological literacy. It embraces global sustainability as the operating premise, and encourages broad-based, systemic thinking throughout. My approach is to provide a solid foundation in ecology and thermodynamics, essential to develop a sense of ecological literacy. From here we use *Stella II*, the computer modeling program used by the authors of *Beyond the Limits* to explore systems dynamics and the prospects for humanity in the 21st century. This presents a great chance to make a transition and take an in-depth look at current environmental issues, beginning with human demographics, which of course is at the core of most sustainability issues.

Daniel Quinn's book *Ishmael* provides an outstanding, provocative look at "how things came to be this way," and opens up excellent discussions on the meaning of worldview. From here we employ the quantitative techniques described in Rees' and Wackernagel's book *Our Ecological Footprint*. The Footprint unit has students calculate their own ecological footprint along with the footprint of the school campus, and to then make their own formal recommendations to both the School Board and City Council on how to reduce human impact on the natural world (both on the campus and within the community).

Finally, the Outdoor Skills and Ethics unit focuses on preparation for a week-long backpacking trip to Zion National Park in southern Utah. Here students learn the art of minimum impact backpacking, and develop a strong appreciation and respect for the natural world. They also build wonderful friendships with their hiking partners, and because it is a backpacking trip, learn directly what "things" they really need and what they can easily do without. It provides them with a direct reading on what is "enough" for them. For many students, this trip is truly the best experience of their high school career.

A snapshot of the year-long course syllabus follows:

Introduction—a look at Easter Island	1 week
The Laws of Thermodynamics	2 weeks
Fundamentals of Ecology	6 weeks
Systems Dynamics and Computer Modeling	3 weeks
Human Demographics	2 weeks
Environmental Issues (could include climate change, ozone loss, biodiversity loss, deforestation, etc.)	4 weeks
Worldview and Environmental Ethics	3 weeks
Our Ecological Footprint	6 weeks
Outdoor Skills and Ethics (with 1 week backpack trip to Zion Nat'l Park)	9 weeks

This course of study results in a broad, holistic perspective on the current state of the world, a sense of hope and empowerment to change things for the better, a wide appreciation and acceptance of the beliefs and opinions of others, and a sense that there is an alternative to "earn a college degree, get the high paying job, work 60 hours per week like your parents, and have no time for yourself or family." Once again, the connecting theme that runs through the entire year is global sustainability.

GOING SCHOOL-WIDE WITH SUSTAINABILITY EDUCATION

It is one thing for a single teacher to implement sustainability into the curriculum, and quite another for an entire school to adopt it as its operating premise. In the fall of 1998, several colleagues at Paso Robles High School and I collaborated to develop a "Week of Sustainability Education" for our students. The week consisted of an opening kick-off assembly that included an introduction to the week's activities, guest speakers, and music. The next four days centered on students rotating to different teachers' classes each day to participate in lessons directly connected to sustainability. Teachers from most departments on campus were represented: Math, Science, Physical Education, Special Education, Social Studies and English. Approximately 1,000 students participated in the event, about half of the student body. Lessons ranged from "The Effect of Raising the Status of Women on Human Population Growth" to "Globalization of Trade and Its Effect on Human Dignity and the Environment." The effort was quite successful (as measured by an end-of-week student questionnaire) at achieving the goal of exposing students to sustainability issues.

In attempting to organize the week, and with the goal of involving as many teachers and students as possible, some barriers presented themselves . . .

1. Most of the teachers who had a desire to participate in the rotation were precluded from joining because they were simply too busy to spend the time required to try something new.
2. Many teachers are too constrained by their curriculum, or by their own belief that there is no room in their program or syllabus to squeeze in anything new—even if they support the effort in spirit.
3. Some teachers felt that the effort to plan the "Week of Sustainability Education" was an attempt by liberals to slip their doctrine into the curriculum. "I'm a conservative, that's

why I believe in conservation" was a phrase heard more than once.

4. Follow-up, even for me (the principal organizer), was difficult due to time constraints.

Still, despite these difficulties, our efforts were successful and rewarding. A special effort was made to obtain administrative support from the outset, and eventually resulted in our Principal supporting our efforts with curriculum pay to plan the rotation. The next step will be to schedule another "Week of Sustainability Education" with as many new colleagues involved as possible.

CONCLUSIONS

It is a complex world—to say the least. How can students connect events in their world? Consider the collapse of the Russian and Indonesian economies, the tremendous hurricanes in Central America, the globalization of trade, tropical and temperate deforestation, the WTO protests, the stock market boom, corporate downsizing, and the collapse of nearly half of the world's major fisheries! Are these to be understood as isolated, disparate events, or is there a larger trend at work here? As a result of their rigorous reductionist training, high school graduates—and many college grads as well—are unprepared to take a systemic view of the world and attempt to understand seemingly unrelated events.

However, studying apparently disparate events in the context of sustainability can provide the adhesive that produces a mental framework to both understand what is happening in the world and to generate ideas to actively create a system in which the prospects for a prosperous economy, and for the ecological health of the biosphere constantly improve. Not only would we be producing graduates capable of earning a good income, but of making the world cleaner, safer, more just and more enduring at the same time.

The ideas proposed in this manuscript are meant to be a starting point. A place from which to begin the process of dialogue to reinvent the way we approach education in this country. It will be an ongoing process—one that will likely never be finalized. It will take vision, commitment and eventually will require the full participation of all staff, students, and citizens. It will be very exciting, provide great hope, energize our students, and furnish the most relevant theme possible to dramatically increase student learning and test scores, and to replace the climate of competition with one of cooperation.

CHAPTER 5

The Role of Higher Education in Sustainable Development Education

Bradley F. Smith

"How many faculty members does it take to change a lightbulb?"

The phrase "paradigm shift" is a very real part of the vernacular of the late 1990's. Even within the title of this book the word "paradigm" surfaces. There is, however, an area of society in which "paradigm shift" should not be used or at best used only sparingly and then only on a case by case basis. That area is higher education.

The phrase paradigm shift invokes thinking outside of traditional boxes or frames of reference. The phrase is intertwined with change in all aspects. Thinking differently, acting differently, and planning differently. A paradigm is a model and a paradigm shift is a model for change. In all of the above, higher education has been and for the most part still is only a weak and limited partner.

The words change and higher education should only rarely be used in the same sentence. Higher education is an institution that is founded and maintained by tradition. Being a member of "The Academy" invokes a fundamental link to the past. The regalia, the ceremonies, the ranking system

Bradley F. Smith, Dean, Huxley College, Western Washington Univ. Bellingham, WA.

and in most instances the attitudes have not changed in centuries. Perhaps one of the most frustrating aspects that inhibits change in higher education is the reward structure. Faculty advance through the maze of academia by adhering to tradition. This is not to fault the faculty member because the system of promotion is already in place and the faculty member desiring promotion is forced to follow a narrow but accepted path. The real downside to this tried, and often less than, true journey is not with the tenured full professor but, rather, with younger junior faculty seeking promotion. It could be argued, unfortunately true in most cases, that it is highly unlikely that a faculty member of some twenty or thirty years is capable of change. After all, such faculty has followed the rules set forth by the Academy and thus been rewarded. The traditional model in academia of faculty who knows a great deal about a narrow subject area, and only communicates with like-minded colleagues, is self-perpetuating. This model "worked" for them and it will "work" for the next generation of faculty. And on and on.

What then of the up and coming junior faculty fresh with the enthusiasm of change and relevance? Higher education seldom rewards enthusiasm, relevance, or change. The prescription at most colleges and universities is clearly defined for promotion and tenure: excellence in teaching (as long as it is discipline relevant), publications (as long as they are discipline relevant), and community involvement (a distant third standard in most cases). When it comes to tenure and promotion, excellence in teaching and scholarship seldom cross disciplines, even less so departments, and definitely not between colleges within a university.

What then about the education of the student? Basically, the Academy educates its students along strict disciplinary lines, to converse with the like-minded and trained and to think theoretically about issues and problems. The defender of the faith would argue that there is nothing wrong with this scenario and that it must work because it has been "working" for a long time. Is there a problem? I believe there is, and it is one that demands attention.

If we educate along disciplinary lines, downplay the need to converse across disciplines and departments and to think primarily in a theoretical mode, for what do we prepare the student that graduates? Let's look at this from another angle. Almost any employer questionnaire will reveal that employers are looking for thinking employees that have a multi-disciplined education, are prepared with problem solving skills and abilities, and are able to communicate. The need for an applied education also surfaces with regularity. Employers want thinking, broadly educated problem solvers. Universities produce narrowly focused single dimensional graduates. There is a problem. The world needs individuals capable of solving puzzles not individuals who can visualize only one piece of the puzzle.

On many fronts, colleges and universities are being challenged with change. The market (employers) is demanding change, the students are demanding change and a rapidly evolving world requires change. Yet, for the most part academia at best has been changing with glacial speed. Education structured around the concept of sustainable development can provide the impetus that forces the agenda of change.

Within the Academy, the initial strands of sustainable development education were found in environmental studies programs. College and university programs in environmental studies began to develop following the emotional and turbulent period of the late 1960s and early 1970s. On the heels of the first Earth Day in 1970 and the passage in the United States of the National Environmental Protection Act, the environment became a visible issue for the first time. Rivers caught fire in Ohio, air pollution caused severe health problems in many cities and this was all covered in color on the nightly news. Environmental studies as an academic field was born. Over the subsequent several decades, environmental studies programs proliferated in academia. With the development of each new program came the inevitable academic question—in what department should the new program be housed? It became clear early on that the Academy was not ready for the concept of inter-disciplinary education. Environmental studies were fine but it had to fit into existing models of administration, i.e. the biology department or the geography department or the political science department. This mind set contradicted what was really needed in terms of education. The problems were broadcast on the news, vivid images of foul air, fish kills, toxic waste dumps and a "dying" Lake Erie. What society needed were problem solvers and it should have been the responsibility of higher education to educate and train such individuals.

In reality, we continued on in the same old way. Environmental studies students housed in biology departments were really biologists and the same was true for other disciplines. Political scientists were just that, as were geographers. Perhaps a course or two were listed as cross disciplinary, however, such offerings were the exception. The outcome was predictable. Graduates of the programs performed, as they were educated. Scientists could not effectively communicate the problems to the public, policy makers did not understand the science behind the problems, and economic considerations were not even on the radar screen. Why, we asked ourselves, were things not getting better?

Ever so gradually, however, change did begin to manifest itself within environmental studies programs. By the late 1980s and early 1990s a maturing process was underway. Many of the programs developed in the early 1970s went through a transformation. Low enrollments during the Reagan years lead to programs being canceled or shifted to minor status rather than an outright major. The initial emotionalism behind the development of environmental studies gave way to the political conservatism of the period. For the most part, there was not considerable dismay on many campuses when environmental studies programs were terminated or reduced in size. Many departments had viewed environmental studies as a drain on their coffers and were relieved when the funding was reallocated along traditional program lines. Environmental studies faculty went back to their traditional specialties and the Academy continued on as before.

The late 80s and early 90s witnessed the beginning of another change in environmental studies on campuses. The environment was once again in the news, this time in a big way. An environmentally besieged Planet Earth was to become Time Magazine's "Person of the Year" in 1989. This was the same year that found medical waste, including spent syringes, washing up on the beaches of North America and the supertanker Exxon Valdez running aground in the pristine waters of Prince William Sound, Alaska. In 1990 the word eco-terrorism entered our vocabulary with the burning of oil fields during the war in the Persian Gulf. Concern about the environment was fast becoming one of the top public policy issues worldwide. This renewed concern towards the environment translated into new environmental studies programs on campuses across North America. Unlike the similar period of the early 1970s, however, this time the programs were being structured along different lines.

From an environmental perspective society had matured over a twenty-year period. Emotionalism was still there but it was now tempered with a degree of pragmatism and fiscal reality. Recycling was fine, as were demonstrations against "polluters," but there must be more than that. Students were asking new questions and demanding new approaches to answering those questions.

What is needed within the Academy is a new and different way for environmental studies programs to be structured. Terms such as inter, multi, and trans have been tacked onto otherwise traditional disciplinary programs for several years. Some worked, others did not. If there is to be a fundamental change in the way students learned about the environment in its broadest sense then there is a need for a new template to be developed. I believe that the issue of sustainability provides such a template.

By its very nature, education structured around sustainable development forces us to dismiss the traditional image of higher education in society. Colleges and universities have been plagued for decades with the "town vs. gown" scenario. Our institutions of higher learning have been viewed as "ivory towers" that were out of touch with the "real" world. They were for better or worse islands within society. Students were taught that their real learning would begin when they entered the "working" world. The traditional university world was seen as removed from the problems of the outside world. Is it any wonder why students were not always in a hurry to graduate or to leave such an environment? After all it was better than "working."

Education structured around sustainable development has the potential to challenge and overturn such images and traditions. The concept of sustainable development forces us to think and hopefully act differently. The traditional discipline and department structured approaches to higher education will not work under such a concept. In addition, colleges and universities will not be able to remain isolated from the greater community they are located in. All of this should not be viewed in a threatening manner but rather as the beginning of needed and I believe healthy change.

The phrase sustainable development is open to many interpretations. The same can be said for education for sustainability. To date, however, there has been some consensus on what the foundations of such an educational model are—education reform, service, and community involvement. While these three areas may not represent the entire spectrum they do provide a point from which to initiate the dialogue.

Education for sustainability cannot be implemented within the classic discipline or departmental framework. It simply does not work. As has been mentioned, the long held model for higher education does not educate in a holistic manner. The problems and concerns of society can not be organized into nicely fitting academic settings or groupings. The issue of sustainable development transcends virtually all of the academic departments within any institution of higher education. Environmental problems, economic vitality, human health, and political realities are only a few of the components within sustainable development. The old model of academic administration does not allow for such a holistic approach. New models, however, are being developed and given a try. Many such models would have been unheard of only a few years ago. One such model or program is the linking of business schools with environmental studies programs. It was not that long ago that you would not even mention the two programs together in the same sentence. Business faculty and environmental studies faculty were at the opposite ends of just about every spectrum. Not so today. In fact, they are actually beginning to understand that they even speak the same language, just with different accents. Waste is the by-product of an inefficient business operation. Waste is pollution. Waste is not our friend. In the end, it's all the same. An economist friend put it another way, "a pollutant should be viewed as a resource out of place." This puts an entirely new twist to the need for pollution control. In that same vein, one can look at the rapid growth of the "environmental industry" sector in the economies of most developed nations. Faculty and administrators are learning that it is not a question of the environment or the economy but a question of both. You cannot have one without the other. Hence there has been a significant growth in the numbers of new business/environment courses and programs in higher education across North America.

The same can be said for other disciplines. There have been over the past few years a growing number of joint programs with environmental studies. Such programs exist with engineering, public health, political science, journalism and law, to name only a few. There are even joint programs with art and theater departments. This trend in combined or joint programs is indicative of students' desire to have an academic foundation and preparation that transcends the traditional confines of the academic structure. There is also a growing body of evidence that such a broadened educational foundation has been beneficial in graduates gaining good positions in their chosen fields. As was mentioned earlier, employers are searching for employees who are capable of thinking outside of a single discipline. Education structured around sustainability will continue to spawn the development of new joint programs and degrees. It is now imperative that the reward systems within the Academy, such as tenure and promotion, encourage faculty to initiate and participate in such programs. This message needs to be echoed loud and clear from the President's, Provost's, and Dean's office of every institution of higher education. Only then will faculty invest their expertise and time in developing such programs.

Having stated the above, it should be mentioned that there are evolving models of multi-disciplined programs in colleges and universities that are successful in initiating education for sustainability. Huxley College of Environmental Studies at Western Washington University is one such example. Unlike most programs in environmental studies, Huxley is a college rather than a department and thus able to have greater flexibility in course design and offerings. The faculty at Huxley range along the academic spectrum from anthropologists to zoologists. Students study from a range of disciplines including political science, economics, geography, sociology, journalism, law and education. This is in addition to solid preparation in the physical and natural sciences such as limnology, toxicology, chemistry and oceanography. Joint programs and degrees have been developed between Huxley College of Environmental Studies and the College of Business and the College of Education in addition to others. Students are required to undertake an off campus internship during their senior year and are encouraged to become active in community programs and activities. The goal within Huxley is to educate and train students to go out and "solve problems." The program is under continual review and change reflecting the rapid changes in the global environmental arena. If employment as environmental professionals is a measure of success, there is solid evidence that the Huxley program is working. Graduates from Huxley enjoy a very high placement rate and degree of satisfaction with their careers. Such data is gathered biannually by in-depth alumni surveys. Perhaps one of the more interesting bits of data is that Huxley graduates are employed equally between the public, private, and non-profit sectors. In addition, a sizable percentage of graduates have become entrepreneurs and started their own businesses during Huxley's thirty-year history.

The second component of needed change in a model of education for sustainability is that of service. Institutions of higher education must provide more than a theoretical education to their students. Students want to be actively involved in real world issues and problems. Volunteerism is a growing component of many programs on campuses across North America. This has been especially true in environmental studies programs. Required internships or fieldwork are a key part of many environmental studies programs. Students benefit in many ways from such hands-on programs. For many students this may be the first time that they apply their education in addressing critical and pressing issues. In addition, an education that involves a service component tends to enhance students' chances when they enter the work force. Valuable connections and contacts is often a by-product of service learning. The third component for learning for sustainability, linking higher education to communities, will be thoroughly covered in a following chapter.

Where then does all this leave us? In the end, I remain optimistic as to the future of education for sustainability. After having bemoaned the fact that higher education is not capable of change, I am encouraged by the changes I have witnessed during the past few years. The traditional walls that have divided the Academy are beginning to crumble. A new foundation for the Academy is being developed, albeit somewhat slowly, and I believe that one of the corner posts of this new foundation is educating for sustainability.

CHAPTER 6

Teacher Education for Sustainability

Two Teacher Education Projects from Asia and the Pacific

John Fien and Rupert Maclean

Our Common Future, the Report of the World Commission on Environment and Development (1987) states that "the world's teachers . . . have a crucial role to play" in helping to bring about "the extensive social changes" needed along the pathway towards a sustainable future (p. xiv). For teachers to play this role successfully they require a commitment to the principles of education for sustainability; without it, they may lack the skills, insights and desire to ensure that their students are provided with opportunities to learn how to contribute to the ways their communities are working to advance the transition to sustainability. The purpose of this chapter is to suggest a rationale for ensuring that education for sustainability is embedded into the initial pre-service and continuing in-service education of teachers. This rationale is supported by principles developed from the experience of two teacher education projects in the Asia-Pacific

John Fien, Faculty of Environmental Sciences, Griffith University, Brisbane, Australia. Rupert Maclean, Chief, Asia-Pacific Centre of Educational Innovation for Development UNESCO, Principal Regional Office for Asia and the Pacific, Bangkok, Thailand.

region, *Teaching for a Sustainable World* (Fien, 1993, 1995) and *Learning for a Sustainable Environment* (Fien, Heck and Ferreira, 1997). Two general points—or caveats—need to be made at the start, however. Firstly, the literature on education for sustainability is relatively new and there has been very little explicitly written on teacher education for sustainability. Thus, it is necessary to draw on literature from the closely related field of environmental education. Secondly, much of this literature is dated and normative, i.e., it is often one or two decades old, and tends to focus on what we ought to do rather than build theory and principles upon accounts drawn from empirical research. A major reason for this may be the long-running situation in which very few teacher education courses have sought to ensure that education for sustainability and/or environmental education are embedded in the professional development experiences of teachers.

THE PRIORITY OF PRIORITIES

The international community has always had an ambitious role for teacher education, perceiving it as "potentially the greatest source of educational change in an organised, orderly society" (UNESCO, 1976). Effective teacher education is a vital first step in producing an environmentally literate population which can, in turn, advance the transition towards sustainability. Not surprisingly, the UNESCO-UNEP International Environmental Education Programme once described the preparation of teachers as "the priority of priorities" for action to improve the effectiveness of environmental education (UNESCO-UNEP, 1990, p. 1). In an early report on this theme, Wilke (1985) stated that:

> The key to successful environmental education is the classroom teacher. If teachers do not have the knowledge, skills and commitment to environmentalise their curriculum, it is unlikely that environmentally literate students will be produced (p. 1).

However, a special commitment from teacher education institutions is necessary to reorient teacher education towards sustainability. This is because education for sustainability requires a new focus and outlook within education which prospective teachers—and teacher educators—may not have experienced in their own education. This new outlook has been described as the exploration of "a new personal and individualised behaviour based on the 'global ethic' which can be realised only through the enlightenment and training of educational professionals" (Simpson et al., 1988, p. 17). Thus, a teacher cannot teach towards sustainability effectively solely by obtaining information on environmental concerns, by studying environmental science. Instead, Simpson et al. argue that:

> Intensive teacher education, not merely orientation, is essential if the present fragmented approaches of traditional education are to be transcended in favour of a holistic, global approach, and inter-disciplinary and a thorough change in both the outlook and preparation of teachers and teacher educators . . . The task is more complex than putting environmental content into existing curricula. (Simpson et al., 1988, p. 17).

Thus, UNESCO's International Congress on Environmental Education and Training, held in Moscow in 1987, resolved that:

> Teacher training is a key factor in the development of environmental education. The application of new environmental education programmes and proper use of teaching materials depends on suitably-trained personnel, as regards both the content and the methods specific to this form of education. Teachers well trained in the contents, methods and process of environmental education development can also play a crucial role in spreading the impact of environmental education at the national level, thus increasing the cost-effectiveness of the efforts made by member States to develop environmental education. . . . There is a need to identify the national objectives of the training of teachers and to develop plans for the training of teachers, which can be implemented by the training authorities. (UNESCO-UNEP, 1988, p. 12)

The United Nations Conference on Environment and Development (UNCED) also highlighted the need for improving teacher education in the expanded field of education for sustainability. In adopting *Agenda 21*, governments committed themselves "to update or prepare strategies aimed at integrating environment and development as a cross-cutting issue into education at all levels within the next three years" (UNESCO-UNEP, 1992, p. 3). Chapter 36 of *Agenda 21* on "Education, Public Awareness and Training" identified training as an important "programme area" specifically called upon educational authorities to assist the development of pre-service and in-service training programmes which address the nature and methods of education for sustainability for all teachers. Similarly, the background paper prepared for the 1997 UNESCO International Conference on Environment and Society: Education and Public Awareness for Sustainability held in Thessaloniki stressed that the messages of education for sustainability "must also be emphasised in pre-service and in-service programmes of teacher training" (UNESCO, 1997, p. 39). More recently still, the Sixth Session of the Commission on Sustainable Development which reviewed the promotion of education for sustainability, called upon national and international action to develop guidelines for reorienting teacher education towards sustainable development (CSD, 1998, p. 8).

ADDRESSING A HISTORY OF NEGLECT

However, while all seem convinced that teacher education has a major role to play in helping teachers address the imperatives of education for sustainable living, we are challenged by an international pattern of neglect in addressing this priority. A claim by Wilke, Peyton and Hungerford (1987, p. 1) that "Few, if any, teacher training programmes adequately prepare teachers to effectively achieve the goals of E.E. in their classrooms" is frequently cited in this regard. While this claim is a decade old, few commentators pause to note that Hungerford and his colleagues were referring to the results a survey conducted in the mid-1970s. This neglect on *our* part may stem from a belief that little has changed since the survey over two decades ago. Certainly, a range of national and international reports in recent years reveal important deficiencies and a lack of co-ordination in the provision of appropriate teacher education for environmental education in many parts of the world (see, e.g. Williams, 1985, 1991; NIER, 1993, 1996).

Fortunately, several projects which address the "priority of priorities" are being developed in various parts of the world. Examples include the *Toolbox* in-service education project conducted by the National Consortium for Environmental Education and Training in the United States (Monroe et al., 1993), WWF sponsored initiatives in the United Kingdon (Bullock et al., 1996), the Environmental Education Initiative in Teacher Education in Europe (Brinkman and Scott, 1994, 1996) and the recently established Consultation Group on Teacher Education within UNESCO's Education for a Sustainable Future Program (UNESCO-EPD, 1998). This paper describes two initiatives in the Asia-Pacific region. The first is an Australian national project called the *Environmental and Development Education Project for Teacher Education*. This project provides resources and professional development for interested teacher educators based upon the *Teaching for a Sustainable World* manual (Fien, 1995). The second is an action research based professional development program for teacher educators in the UNESCO Asia-Pacific region, called *Learning for a Sustainable Environment – Innovations in Teacher Education.* These case studies were selected because they illustrate an important lesson about the dissemination and adoption of innovations in environmental education. This lesson comes from research on curriculum innovation and professional development for environmental education at the school level, but the lesson is equally applicable at the level of teacher education. The lesson is that there is a strong ecological kind of relationship between curriculum development, professional development, practitioner-based research and networking. They are so interconnected and interdependent that none can survive without the other.

Hart (1990) has used the concept of the "reflective practitioner" equipped with skills for "reflection-in-action" (Schon, 1983) to describe this "ecological" approach to professional development for environmental education. He argues that the constructivist epistemology and the focus on interdependence (of teaching practices and contexts and between teaching colleagues and their communities) in this approach is philosophically consistent with an ecological world view and, thus, also with the objectives and guiding principles of environmental education. Hart has identified a range of parallels between ecology and educating teachers environmentally, as reflective practitioners:

A reflection-in-action perspective on teacher education places emphasis on educational aims and consequences as well as the technical skills of teaching. Teachers [and teacher educators] are encouraged to consider ethical and value-based issues and this facilitates environmental education goals. Teachers [and teacher educators] are also encouraged to contribute to the formulation of policy at classroom, local, and national levels thus acknowledging the political nature of human interpretations of the ecological process of change. . . .

Teacher education programs based on a reflection-in-action paradigm emphasize a process model of education where teachers [and teacher educators] . . . monitor and evaluate their own practice reflexively, that is, an action research model, a cyclical process in which teacher action-reflection-improved action is seen as a dialectic between theory and practice, much like the principle of reciprocal relationships is viewed in ecology.

A reflection-in-action approach to teacher education would not propose to eliminate or replace educational disciplines but rather to use them to emphasize relationships to the teacher's own thinking about practice. That is, the traditional subjects, when combined with classroom experiences, would provide a basis on which to develop teachers' personal theories of action. The associated ecological principle is that of optimization.

Within this new perspective for teacher education, the effective teacher is not one programmed with research (theory)—based answers to many discrete teaching situations. Rather, the reflective teacher [and teacher educator] is one who is able to conceive his/her own teaching in purposeful terms—one who is able to size up a particular teaching situation, choose an appropriate action, judge results in relation to context and original purposes. This concept operationalized is congruent with self regulation within ecological systems. (Hart, 1990, pp. 14–15)

This perspective on teacher education has led to a sustained critique of centrally-driven, technicist "research and development" models of educational change in the Asia-Pacific and other parts of the world. As

Robottom (1989) has argued, such technicist approaches reduce the role of teachers [and, in our case, teacher educators] to that of "passive receivers" of centrally produced curriculum materials who can be regarded as "technicians" applying the ideas of their "professionals" (p. 441). Robottom concludes that two key characteristics of environmental education are undermined in the "research and development" approach: the development of skills for critical thinking, self-evaluation and reflection, and the need for knowledge, values and actions of participants to be engaged in the process of change. Just as we would not want to neglect these objectives and principles in our teaching, we should not neglect them in our professional development work. The two projects described in this paper have sought to maintain the ecological relationship between curriculum development, professional development and practitioner-based research to the extent that available resources allow.

TEACHING FOR A SUSTAINABLE WORLD

Teaching for a Sustainable World is a revised, expanded international edition of the materials developed in the Australian Environmental and Development Education Project for Teacher Education, and contains twenty-six 3–5 hour workshop modules on education for sustainability for use in teacher education. These modules were written by twenty-four educators from twenty-one different universities, government departments and NGOs across Australia, New Zealand, the UK, Nepal and South Africa who worked co-operatively to write, test, and revise the modules. The original Environmental and Development Education Project for Teacher Education was a joint initiative of Griffith University, the Australian Association for Environmental Education and AusAID, the Australian international development agency, and arose from a perceived need to support teacher educators in responding to the challenges of integrating environmental and development education following the Earth Summit in 1992. An international review and collaboration process then redeveloped this Australian work into a twenty-six module teacher education guide entitled *Teaching for a Sustainable World* (Fien, 1995).

The Australian Experience

The themes of modules in the initial Australian work were selected after an examination of the research and teaching interests of interested teacher educators. The themes they suggested were then reviewed by the project management team in the light of the literature on sustainable development and environmental education. Each participant was then asked to write a 3–5 hour workshop suitable for use in pre- and in-service teacher education courses. A common but flexible set of headings were provided to guide in the format of the workshop and a circle of "critical friends" provided for each author. Copies of first drafts were circulated to the critical friends who reviewed and trialed all or part of the workshop. Second drafts were submitted to critical friends also and then to the project office for final editing. Dissemination workshops were held for all teacher education schools across Australia during 1993 and 1994 while the Australian Association for Environmental Education featured "teaching for a sustainable world" in its national Professional Development Programme between 1994 and 1997 (see Smith, 1997).

The International Experience

Noting the success of the Australian program, the Environmental Education and Training Unit of UNEP commissioned an international edition of the Australian work under the auspices of the UNESCO-UNEP International Environmental Education Programme. A consultation workshop was conducted as part of an Asia-Pacific regional seminar on environmental education and teacher education (Griffith University 1993). This workshop made several recommendations for the international adaptation, including: (i) replace any overtly Australian examples except within particular contexts; (ii) increase the range of case studies to give a more balanced coverage of regions of the world; (iii) provide particular examples and suggest support for the use of the manual in developing countries; and (iv) integrate where practicable the themes of culture and religion in relation to sustainable living; sustainable agriculture; alternative technology; the impact of rapid industrialisation and urbanisation on living in an ecologically sustainable way; and health and environment in relation to choosing healthy ecologically sustainable lifestyles. Associate editors were appointed from the UNESCO-UNEP International Environmental Education Programme and an international reference group established to act as critical friends in the development of new modules and adaptation of the original ones.

In all, twenty-six modules were written to illustrate how environmental and development themes are related, and to provide practical assistance for teacher educators who would like to include these important themes in their courses. The titles of the modules are listed in Table 1. Modules 1–4 provide an overview of the interdependence of environment and development, and introduce environmental and development education and the linkages and similarities between them. These four workshops

Table 1. Workshop modules in *Teaching for a Sustainable World*

Introductory Workshops
1. A View of a Sustainable World
2. Environmental Education
3. Development Education
4. Development and Environmental Education Exploring the Links

Modules on the Nature of Sustainable Living
5. Teaching for Ecologically Sustainable Development
6. Introducing Sustainable Futures
7. Appreciating Sustainable Futures
8. Culture and Religion: Important Lessons for Sustainable Living
9. Exploring Environmental Beliefs

Modules for Subjects, Curriculum Themes and Topics
10. New Science: A New Worldview
11. Health, Environment and Community Development
12. Environmental and Health Education in Rural Communities
13. Community Action for Sustainable Development
14. Community-based Environmental Education
15. River Studies for Primary Schools
16. Consuming for Sustainability
17. Women, Environment and Development
18. The Population-Food Debate
19. Sustainable Agriculture and Rural Development
20. Traveller or Tourist: Tourism in the Developing World
21. Hope or Despair: Sustainable Living in Informal Settlements
22. Waste Management and Future Problem-Solving
23. Alternative Technology
24. Refugees and Development

Concluding Modules
25. Analyzing Educational Resources for Environmental and Development Education
26. Personal Power and Planetary Survival

may be considered as a hub and the remaining ones as spokes that address particular themes and needs in education for sustainability. Modules 5–9 explore the nature of sustainable development, sustainable futures and the role of environmental values and beliefs in culture and religion. They may be used in any combination and sequence but, preferably, after the four core workshops have been completed. Modules 10–24 provide workshops on how the theme of sustainable living may be introduced into the teaching of particular subjects, such as science (Module 10), health (Modules 11 and 12) and consumer education (Module 16), and in the teaching of particular cross-disciplinary themes, such as community based environmental education (Modules 13 and 14), women, environment and development (Module

17), population, food and agriculture (Modules 18 and 19), and waste management (Module 22). The focus of these modules is to illustrate ways in which small starts may be made to change the focus of traditional subjects and topics to begin the process of education for sustainability. Module 25 provides guidelines for reviewing existing curriculum materials to diagnose how they may require revision or supplementation in order to ensure that resources are conducive to education for sustainability. Once again, the purpose of this module is to provide a starting point for those who may lack up-to-date curriculum materials and otherwise feel disempowered from beginning the process of education for sustainability. Module 26 provides a range of strategies for aiding teachers and student teachers to clarify their sense of commitment to sustainability and uncover the inner resources they have to feel empowered to teach for a sustainable world.

Using Teaching for a Sustainable World

The audience for *Teaching for a Sustainable World* includes universities, teachers' colleges, curriculum development centres, education systems, professional associations, teachers' centres, community environment and development organisations and schools. Each of the workshops has been written with the needs of workshop facilitators in mind with the activities phrased in terms of the things that workshop facilitators needs to consider doing when planning and leading a workshop. Generally, each module is based upon a concrete learning experience which requires participants to work individually, in pairs, or small groups to complete a task. Thus, the workshops promote active experiential approaches to learning and model the sorts of learning experiences that can achieve the wide range of knowledge, skill, values and participation objectives of environmental education. Lecture-style input is kept to a minimum and always referred to as a "mini-lecture." Facilitators have been strongly encouraged to obtain or develop local variations of the materials and to adapt them so that their workshops are as relevant as possible to the cultural and educational contexts in which they are working. Indeed, locally adapted versions and translations of *Teaching for a Sustainable World*, in whole or in part, have been encouraged, with such efforts underway in Thailand, Greece, South Africa, and Taiwan.

Unfortunately, the missing element in this work has been the practitioner-research focus recommended by Hart. Structures were not established to encourage the original authors and new participants to adopt a research stance on their use of *Teaching for a Sustainable World*. This had the effect that few networks of teacher educators have been established to monitor and encourage the on-going implementation of the innovation.

This weakness in the ecology of the *Teaching for a Sustainable World* was addressed in the development of the second project.

THE LEARNING FOR A SUSTAINABLE ENVIRONMENT—INNOVATIONS IN TEACHER EDUCATION PROJECT[1]

The *Learning for a Sustainable Environment—Innovations in Teacher Education Project* began in the first half of 1994 as a joint initiative of UNESCO's Asia-Pacific Centre of Educational Innovation for Development (ACEID) and Griffith University in Australia. The primary goal of the project has been to assist teacher educators in the Asia-Pacific region to include the educational purposes and innovative teaching and learning strategies of education for sustainability in their programmes.

Development of the Project

Four regional and subregional gatherings were held leading up to the *Learning for a Sustainable Environment* project. The first was an International Experts Meeting on *Overcoming the Barriers to the Successful Implementation of Environmental Education through Teacher Education* held at Griffith University in Brisbane 4–9 July 1993. There, delegates and official observers from seventeen countries developed guidelines and recommendations for achieving the goal, embodied in this title of the meeting. Next, a seminar on *Environmental Education and Teacher Education in Asia and the Pacific* was held in Tokyo, Japan on 20 October to 5 November 1993. This meeting's major recommendation called for the preparation of prototype teacher education materials for trialing and revision followed by region-wide implementation. This recommendation was in accord with the focus on competencies for the environmentally educated teacher identified in the Brisbane meeting. The recommendation was also adopted by the South East Asia Ministers of Education Organisation (SEAMEO) and UNESCO subregional conference on environment education and secondary teacher education held in Penang, Malaysia, 6–9 December 1993. Finally, a meeting was hosted by Griffith University in Brisbane from 26 June to 1 July 1994 to plan a specific project to promote innovations in environmental education in teacher education, *Learning for a Sustainable Environment—Innovations in Teacher Education.*

[1] The description of this project is based upon the draft report of the Planning Group Meeting held in July 1994. The experience and scholarship of all who attended this meeting is reflected in the ideas in this section.

Based on country reports submitted to these UNESCO seminars, a series of guiding principles aimed at enhancing the quality and relevance of environmental education were consolidated at the final meeting as a summary of best practice in contemporary environmental education in the region:

- Environmental education considers the environment in its totality, i.e., ecological, political, natural, technological, sociological, aesthetic, and built environments.
- Environmental education develops awareness of the importance, beauty and wonder that is, and can be, found in these aspects of the environment.
- Environmental education explores not only the physical qualities of the human relationship with the environment, but also the spiritual aspect of this relationship.
- Environmental education is a response to the challenge of moving towards an ecologically and socially sustainable world.
- Environmental education is concerned with the interaction between the quality of the biophysical environment and the socio-economic environment.
- Environmental education transcends the division of knowledge, skills and attitudes by seeking commitment to action in an informed manner to realistic sustainability.
- Environmental education recognizes the value of local knowledge, practices, and perceptions in enhancing sustainability.
- Environmental education supports relevant education by focusing learning on local environments.
- Environmental education considers the global as well as the local environment. Since the world is a set of inter-related systems, there is a need for a world perspective on environmental issues.
- Environmental education focuses on current and future perspectives on environmental conditions.
- Environmental education is interdisciplinary and can be taught through and used to enhance all subjects in the curriculum.
- Environmental education emphasises participation in preventing and solving environmental problems and revokes the passive accumulation of information about the environment. It should foster and arouse a sense of personal responsibility and greater motivation and commitment towards the resolution of the environmental situation.

- Action is both a vehicle for and an outcome of environmental education.
- Environmental education develops the skills:
  - to identify alternative solutions for the environmental situation;
  - to clarify the values associated with the alternatives;
  - to use these values to make decisions about which alternatives to choose.

Developing educational programs based upon those principles can pose problems for many teachers, especially those who work in formal, centrally organized education systems. As Peter Fensham stated in the Final Report of this Planning Group Meeting:

> The curriculum goals of environmental education overlap with, but also differ quite markedly from, those for other more familiar components in the school curriculum . . . Environmental education in its fullness involves very major changes in the ways teachers conceive of, and act in their classrooms. There are many ways teachers can contribute to education for the environment that involve smaller, but significant changes of thought and action. Conceptual and real behavioural changes, however large or small, are not easy and teachers are no different from others in not finding significant change easy.
>
> Invitations to change, or to try innovatory teaching strategies are almost inevitably seen as "additions" and hence requiring extra time and effort. The suggestion to innovate often comes as part of an external innovator's timetable and not at the point in the teachers' lives when they are dissatisfied with their present practice, and hence are looking for alternatives to solve a problem they personally recognise. The uncertain outcomes of using alternative pedagogies also are more likely to be seen as threatening the teacher's authority and stability of their classrooms than as improving these relationships, as they may in fact do. (UNESCO-ACEID, 1994)

The Professional Development Process

The environmental education of teachers must address problems such as those and, to the greatest extent possible, be consistent with the principles of environmental education. The problem encountered by those who were developing the *Learning for a Sustainable Environment* project was to find a way of supporting this in a culturally diverse region that includes both the most populous country in the world (China) and some of the smallest (Pacific island states) as well as some of the wealthiest and some of the poorest. A standard set of prototype materials could not be culturally or educationally relevant nor could they be the teacher educator's own. A solution to this problem was found in Hart's (1990) and Robottom's (1989) conceptions of professional development for environmental education outlined above. This perspective on teacher education led to the adoption of a set of key principles for ensuring an ecological relationship between curriculum, development, professional development and research in the project. These principles included:

- The project's emphasis should be on personal and professional development of teacher educators rather than on the production of resources.
- Collegial and collaborative approaches underlie successful professional development. Hence, a system of critical friends should be established to review and trial early drafts of all prototype materials and to advise on their development.
- Active participation and critical reflection are essential components of professional development. Participation in these workshops will provide such opportunities and assist teacher educators to:
  - clarify the strengths and limitations of their present practice;
  - establish their personal ownership of the project through the view, critique, revision, adaptation, and curriculum resources.
- The cultural and educational diversity in the region requires a framework for professional and curriculum/resource development which gives direction to participants but is flexible enough to accommodate local concerns and priorities.
- Existing networks in the region must be used and strengthened to facilitate the diffusion and dissemination of the innovative approaches developed by participants.
- The prime focus of the evaluation of the project should be on the teacher educators and the quality of their learning in relation to their use of innovatory teaching and learning strategies.

Overview of Project Strategy and Activities

These principles of environmental education and professional development were incorporated into a three stage strategy as illustrated in Table 2. Stage 1 of the project began in June 1994 and involved six countries—the Philippines, Australia, Papua New Guinea, New Zealand, Hong Kong, and Fiji. Several teacher educators from these countries wrote a set of workshop modules, each about 3–6 hours in duration, for use at either the inservice or preservice education level. A network of critical friends was established for each Stage 1 author from the attendance list of the 1993

UNESCO Meetings; and over one hundred teacher educators from over twenty countries have provided advice and suggested revisions. The Network for Environmental Training at the Tertiary Level in Asia and the Pacific (NETTLAP) located within the UNEP Regional Office for Asia and the Pacific in Bangkok also sponsored a feasibility study and survey to establish wider perceptions of the need for the project and develop a mailing list of persons in the region interested in participating in the project. The guidance provided to these authors enabled them to write modules that reflected the following principles:

- The modules were *written by authors from their own use* but in a sufficiently flexible way that they can be shared with colleagues in their own country and other parts of the region.
- All modules were to be written in a *workshop format.*
- The modules were based on *active adult learning* and *experiential* learning processes which involved direct personal experience and reflection in order to identify, develop and reinforce new concepts and skills.
- The modules were *not to be seen as a finished product* but reflected a dynamic resource that can be constantly evolving as it is analysed, adapted, trialed and evaluated in different contexts. This reflects a concern for professional development as opposed to product development, for the open-ended resource would allow for possible addition of local and/or self-developed materials and adaptation.

The draft modules were reviewed both by colleagues in the authors' own countries and a network of over 100 "critical friends" across the region, and revised accordingly. While the modules were written primarily for use of the authors in their own countries, the network of critical friends endorsed all of the modules as being suitable for possible adaptation and use in other countries of the region. Nine workshop modules were developed and trialed and, on the basis of feedback from the trials in Stage 2, were restructured and edited into ten modules on the topics listed in Table 2.

Stage 2 of the project involved professional development workshops in Thailand (1995) and Australia (1996) and included participants from the Stage 1 countries and a widening circle of network members from Malaysia, Indonesia, Vietnam, India, the Philippines, Korea, Singapore, Japan, and Thailand. All Stage 2 participants agreed to trial at least one—and often more—of the Stage 1 modules after an evaluation training workshop and with continued guidance from the project office. The aim of these trials and evaluations were twofold: to provide guidance on ways of revising the

Table 2. The process of progressive improvement of versions of the modules through review and reflection

Stage	Module Version	Review and Reflection
	1993 Seminars and Conferences	
1	1994 Workshop (Crystal Creek	Project Principles and Strategy
	Draft 1 Modules	Comments by Critical Friends
	Draft 2 Modules	1995 Workshop (Pattaya)
	Trial Modules	Return to Authors
	Version 1	
2		In-Country Analysis and Adaptations
	Trial Modules (country)	
	Version 2	
	Trial Modules (country)	
	Version 3	
	Trial Modules (country)	
	Version X	
		Trial and Evaluation
		Action Research Reports
	1996 Review and Evaluation Workshop (Tokyo)	
3	Publication	Original Author and Editorial Panel
	Version	Further In-Country
		Analysis and Adaptations

Table 3. Workshop Modules in *Learning for a Sustainable Environment*

1. Education for Sustainability
2. A Whole School Approach to Environmental Education
3. Experiential Learning for the Environment
4. Storytelling for the Environment
5. Indigenous Knowledge for the Environment
6. Values Education for the Environment
7. Enquiry Learning and Teaching for the Environment
8. Using the Environment as a Resource for Learning
9. Community Problem Solving
10. Appropriate Assessment for Environmental Education

modules for use in different contexts, and to assist teacher educators to identify the strengths and limitations of their present teaching practices and to develop professionally through the reflection and evaluation activities in action research process. Participants were supported by a *Review Guide* which assisted with the four trial tasks:

- to analyze the Stage 1 module(s) for local cultural and educational relevance;
- to adapt/rewrite the module(s) as a result of the analysis;
- to teach a module(s) and gather data to use in evaluating it (them);
- to write a report on the trial and results.

This action research network process was designed to cater to the professional development goals of the project and involved four steps:

1. *Analysis*: The working principles for the development of modules stressed that the modules were to be written "by authors for their own use". Each original author considered his or her own professional context when preparing a module. Trialists were encouraged to take similar considerations and to analyse the materials to identify their relevance—or otherwise—to local cultural, geographical and educational circumstances.
2. *Adaptation*: The results of this analysis were used by trialists to help adapt the modules to meet the needs of their own students in their own countries. This included: deleting or changing certain activities and resources, incorporating their own resources and activities, translating overhead transparencies and readings, etc.
3. *Trial and Evaluation*: The adapted modules were trialed and evaluated in a range of settings—with pre-service teacher education students, with experienced teachers attending in-service workshops, and in workshops attended by teacher educators as well as local government and NGO officers.
4. *Reporting*: When the participants had completed the trial and evaluation of their modules, they wrote a report in which they reflected on both the quality of the learning experience provided for participants and on the value of the trial/action research process for their own professional development.

The Project Office prepared a Review Guide to provide participants with structured professional development experiences during the analysis, adaptation, trialing and reporting phases of their work as they sought to answer three key questions:

1. What has this workshop done for my personal professional development?

2. How and for what reasons are the adaptations needed?
3. Was the adaptation and trial an effective experience for me?

Stages 1 and 2 of the project were evaluated at a seminar in late 1996 which was sponsored by the National Institute for Educational Research (NIER), in Japan. This provided a forum for participants to present their action research reports, to evaluate the action research approach to professional development and to plan the dissemination activities of Stage 3 of the project which began in 1997.

Influences on the Project

In evaluating the action research network approach to professional development, all network members indicated that they gained a great deal from the action research process of reviewing, adapting, trailing and evaluating the project modules. Many indicated they were transformed from being an *environmentalist* into an *environmental educator* through this process. They stated that they developed a rich knowledge and understanding of the scope and purposes of environmental education during the project, as well as, the ability to choose and make use of the most appropriate teaching and learning strategies to develop the environmental education competencies of student teachers and teachers attending inservice education workshops. Of importance to us in this chapter is their analysis of the influences that they believed contributed to this high relative degree of success of the project. A focus group interview approach was to used to identify and clarify these influences.

Firstly, they agreed that the professional development and practitioner research process was more personally, professionally and culturally relevant to them and their institutions than previous product or materials focussed projects they have encountered. Many participants stated that they developed an increased sense of confidence and personal efficacy not only through using the innovative teaching methods but by being acknowledged as researchers. Many commented that the "training" they received in cultural analysis, data gathering and report writing greatly aided their development as researchers. They identified the assistance provided by the project office in these tasks as invaluable, especially for those who were relatively new to the research process. Researching one's own practice was seen as a very relevant and accessible approach to research for inexperienced educators. In this way the project was described as an important contribution to helping develop a culture of research among teacher educators in the Asia Pacific region.

Secondly, participants commented on the positive support they had received from their colleagues and institutions. This came variously in the form of encouragement, assistance with secretarial support, photocopying, etc. and, in many cases, from the direct relevance to existing college courses and subject content of their research. Other participants were able to incorporate their involvement in the project into the in-service education outreach activities of their institutions or professional associations. These factors meant that the project often could be seen as part of the normal scope of duties of participants. However, the commitment of many participants meant that they devoted a considerable amount of their own time (and sometimes funds) in revising, translating, adopting and delivering workshops and the evaluation of their teaching.

Thirdly, participants commented on the collaborative processes adopted in this project and the role of the project office in facilitating collaboration between network members from countries. In particular, they noted the sharing of ideas via newsletters and reports, the training workshops and the flexible guidelines which addressed professional development needs at different stages in the project. The guidelines provided by the project office for evaluating and reporting on the progress of the trials at subsequent workshops were seen as a key influence on the high level of professional development experienced by participants. The series of regular (roughly annual) seminars also were seen as important influences for several reasons. They provided opportunities for obtaining the advice and support of experienced colleagues from other countries. In addition, attendance at international conferences was motivating for participants as it both ensured that they met the timelines for all tasks and gave increased status to their work in EE in teacher education.

Another value of the collaborative nature of the project which was identified, was the lack of ownership or control of the project by any one group. This came about from the multiple sources of funding for the project and the dependence of the project office on the contributions from network members. Also, the various iterations of the project modules as they were drafted, reviewed, analysed, revised, evaluated and again revised in different forms in different countries meant that all participants had opportunities to contribute to the products of the project—and to be seen as an "expert." This helped to provide an atmosphere of egalitarian co-operation amongst network members and might be seen to represent a more democratic approach to international development assistance in education than approaches based upon top-down expert-novice relationships.

These five factors have contributed to a sustained research effort in teacher education by environmental educators in just under twenty countries. At the time of writing, Stage 3 of the project is underway. This involves participants from the earlier stages taking the role of national coordinators for the establishment of action research networks of teacher educators in their own countries. National Coordinators have been appointed and two training workshops held. As a result of these, national workshops were held in Thailand, Indonesia, Vietnam, and the Philippines in the first half of 1998 and national action research networks formed to parallel the regional one. An important research question presently being considered is whether the five factors which sustained the international research effort are influential in sustaining these national processes or whether other factors encourage and/or constrain the activities of national action research networks in teacher education for sustainability.

CONCLUSION

The two projects described in this chapter have invited teacher educators to consider the imperatives of education for sustainability, to critique, trial and share their work, and to interact with members of the international and national community in order to find ways of addressing the global crisis of environmental and sustainability education. The projects have illustrated the value of principles of professional development recommended by Hart and Robottom to argue the case for teacher educators to ground their work in an ecological approach which integrates curriculum development, professional development and reflective practitioner-based research. The neglect of the reflection and research element in the ecology of *Teaching for a Sustainable World* made its dissemination and adoption problematical despite the collaborative processes of curriculum development involved and the excellent materials that have been produced. It is the hope of all those involved in the *Learning for a Sustainable Environment—Innovations in Teacher Education Project* that the emphasis on all three elements in the ecology of professional development, curriculum development and practitioner research will contribute to the ongoing success of this international project and its national dissemination activities. Perhaps, then, as more of us are working this way, as teacher educators around the world, we will be able to stand with those who have refused to "stand aloof from the decisions about how and whether life will be lived in the twenty-first century" (Orr, 1992, p. 145).

REFERENCES

Brinkman, F.G. and Scott, W.A.H. (1994) *Environmental Education into Initial Teacher Education in Europe (EEITE): the state of the art*, ATEE Cahiers No. 8, Brussels: Association of Teacher Education in Europe.

Brinkman, F.G. and Scott, W.A.H. (1996) "Reviewing a European union initiative on environmental education within programmes of pre-service teacher education", in *Environmental Education Research*, 2 (1), Bath: Carfax.

Bullock, K.M., English, T., Oulton, C.R. and Scott, W.A.H. (1996) "Reflections on an environmental education staff development initiative for teacher educators", in Champain, P. and Inman S. (eds.) *Thinking Futures: Making Space for Environmental Education in ITE—A Handbook for Educators*, Godalming: WWF.

CSD (1998)*Transfer of Environmentally Sound Technology, Capacity-building, Education and Public Awareness, and Science for Sustainable Development*, Report on Sixth Session,: Commission on Sustainable Development.

Fien, J., ed. (1993) *Teaching for a Sustainable World: The Environmental and Development Education Project for Teacher Education*, Brisbane: Australian Association for Environmental Education, Griffith University and Australian International Development Assistance Bureau.

Fien, J., ed. (1995) *Teaching for a Sustainable World*. The UNESCO-UNEP International Environmental Education Programme for Teacher Education. Brisbane: Griffith University.

Fien, J., Heck, D. and Ferreira, J. (1997) *Learning For A Sustainable Environment*, Brisbane: Griffith University.

Griffith University (1993) *The Final Report: UNESCO Asia-Pacific Regional Experts' Meeting on Overcoming the Barriers to Environmental Education through Teacher Education*, Brisbane: Griffith University.

Hart, P. (1990) "Rethinking Teacher Education Environmentally", in *Monographs In Environmental Education and Environmental Studies*, Vol. VI, Troy, Ohio: North American Association for Environmental Education.

Monroe, M. et al. (1993) *Toolbox Series 1–5*, Ann Arbor: NCEET.

National Institute for Educational Research (1993) *The Final Report of a Regional Seminar: Environmental Education and Teacher Education in Asia and the Pacific*. Tokyo: National Institute for Educational Research.

National Institute for Educational Research (1996) *The Final Report of a Regional Seminar: Learning for a Sustainable Environment: Environmental Education in Teacher Education*. Tokyo: National Institute for Educational Research.

Orr, D. (1992) *Ecological Literacy: Education and the Transition to a Postmodern World*, Albany: State University of New York Press.

Robottom, I. (1989) "Social Critique or Social Control: Some Problems for Evaluation in Environmental Education", *Journal of Research in Science Teaching*, Vol. 26, No. 5, pp. 435–443.

Schon, D. (1983) *The reflective practitioner: How professionals think in action*, New York: Basic Books.

Smith, J.H. (1997) "The national professional development program for environmental education in Australia", in *International Research in Geographical and Environmental Education*, 6 (3), New York: Channel View Books/Multilingual Matters Ltd.

UNCED (1992) "Promoting Education and Public Awareness and Training", *Agenda 21*, United Nations Conference on Environment and Development, Conches, Ch. 36.

UNESCO (1978) *The Final Report: International Conference on Environmental Education*, Paris: UNESCO.

UNESCO (1980) *Environmental Education in the Light of the Tbilisi Conference*, Paris: UNESCO.

UNESCO-ACEID and Griffith University (1994) *Final Report of the Planning Group Meeting for the UNESCO-ACEID Project, Learning for a Sustainable Environment—Innovations in Teacher Education*. Brisbane: Griffith University.

UNESCO-EPD (1997) *Educating for a Sustainable Future: A Transdisciplinary Vision for Concerted Action*, Thessaloniki: UNESCO.

UNESCO-EPD (1998) *Report of Advisory Group on Teacher Education for Sustainability*, Thessaloniki: UNESCO.

UNESCO-UNEP (1988) *International Strategy for Action in the Field of Environmental Education and Training for the 1990s*, Paris: UNESCO and Nairobi: UNEP.

UNESCO-UNEP (1990) "Environmentally Educated Teachers: The Priority of Priorities", *Connect*, Vol. XV, No. 1, 1–3.

Wilke, R.J., Peyton, R.B. and Hungerford, H.R. (1987) *Strategies for the Training of Teachers in Environmental Education*, UNESCO-UNEP International Environmental Education Programme, Environmental Education Series No. 25, Paris: UNESCO Division of Science, Technical and Environmental Education.

Williams, R. (1985) *Environmental Education and Teacher Education Project*, Report to the Educational Advisory Committee of the World Wildlife Fund (UK), University of Sussex.

Williams, R. (1991) *Teacher Education Survey: Environmental Education: Interim Report*. Education Network for the Environment and Development, Unpublished.

World Commission on Environment and Development (1987) *Our Common Future*, Oxford: Oxford University Press.

ADDITIONAL REFERENCES

Fien, J., ed. (1993) *Environmental Education: A Pathway to Sustainability?* Geelong: Deakin University Press.

Fien, J., ed. (1993) *Education for the Environment: Critical Curriculum Theorising and Environmental Education*, Geelong: Deakin University Press.

Fien, J. and Tilbury, D. (1996) *Learning for a Sustainable Environment: A New Agenda for Teacher Education*. Bangkok: UNESCO-ACEID.

CHAPTER 7

Education for Understanding Science and the Earth System for Long-Term Sustainability

M. Patricia Morse

INTRODUCTION

Within our human society, we are facing a challenge of complex interacting problems that has the potential to affect us in our lifetime, and will certainly determine the quality of life for our children in the future. It is now clear that human activity affects our global environment in ways that we did not think or imagine as being possible only a few years ago. It appears that these impacts will affect the general quality of life, and more specifically, the availability of energy, minerals, and other natural resources, as well as the quality of the air we breathe and the water we drink. How is it that this human species, with its ever-expanding complexity of technological innovations, its great powers of discovery, and its new developments in the exchange of knowledge finds itself attempting to understand these interactions between our many human systems and our physical, biological, and geological natural systems? A consensus is building that we must take steps to not only understand this system but develop new life patterns that will

M. Patricia Morse, Professor of Biology, University of Washington, Seattle, WA.

assure we and our future progeny will interact, eco-evolve and integrate our activities for a sustainable world.

Working toward this sustainable world means we need to understand all the elements and parts of this global system, and begin to design decisions that will lead to a more balanced system of change. We need science, society, and technology as central factors to the solutions both on a global decision-making scale and in our everyday lives. Clayton and Radcliffe[1] define the connections between system theory and the earth systems by defining the world as a very complex system which contains complex subsystems. These subsystems include various ecological and biological systems, weather systems, and multiple human, social, and economic systems. They continue to note that as the earth system evolves over time, the environmental factors that shape the outcome can change in a variety of dimensions. Of these changes, some will not be significant whereas others can have significant impact. We need to avoid irreversible situations, especially if human effects make important and potentially renewable resources extinct. Monitoring these dimensions is essential to developing appropriate responses to danger signals. General systems theory provides a unifying framework where explanatory and analytical tools will help work toward good decisions. A logical approach is to model the Earth's systems. By continuously looking for the underlying principles common to the various parts of the Earth's systems, we can begin to address possible solutions toward living in harmony and balance within this biosphere.

Lubchenco[2] notes "the new and unmet needs of society include more comprehensive information, understanding and technologies for society to move toward a more sustainable biosphere, one which is ecologically sound, economically feasible and socially just." To vision such a biosphere for our children, educating for sustainability is everyone's business!

BACKGROUND

In the 1990s, science exploration has led us into significantly expanding and deepening units of understanding from universal visualizations of the earth and planets, to the very minute building blocks of the Universe. Scientific knowledge, as a way of knowing about the natural world, has been utilized in endless ways, by the discoverers, humankind. Now it is incumbent on us to know how we can develop ways of using this knowledge that will address sustainability of the natural world environments around us, as we know it, for future generations. Life time patterns of education in how and why science is important for understanding the complexity of sustainability is an exciting challenge.

Science is a human endeavor. A species of this planet, *Homo sapiens*, humankind is endowed with a complex nervous system, including a brain that allows us to record history, develop wonderful innovations through design, research endless components of this complex world in which we live, and express our understandings of our environments through art, literature, and other human endeavors. We are able to form endless assemblages of thoughts, words and visualizations to synthesize and philosophize about our surroundings. We have the remarkable ability to plan for the future through this synthesis, through this knowledge, and through the techniques of prediction based on current understandings. Science holds an attraction for all of us. Yet in the past, scientists and others have described, pictured and assigned science as a complex endeavor, only practiced and understood by an elite, and often passed over in synthesis activities whether at the decision tables of governments or in the policy making institutes of the future. But things are changing. As our knowledge base grows, large numbers of scientists are coming to the tables and joining the conversations and meeting the obligation to the public to provide new ways of thinking. Synthesis in science is becoming a household word; educational issues now share the stage with research; scholarly works and pronouncements for the future of our planet involve the practicing scientists in the deliberations, delegations, writings and actions. The words environment and sustainability are now in the forefront of planning, in the corporate, government and private worlds of citizens, both here in the US and around the World.

The trends of science and our historic relations to them, have been eloquently translated for us by such authors as Wilson[3] whose newest book, "Consilience" holds new ideas about the richness in synthesis and understandings and our historic interactions with the Earth system. This, and the myriad of accounts from national and international leadership groups, leads us to understand that we are developing a new trend, one toward unification and synthesis of information to solve the numerous problems that face us within our global environment. We need ways to assure these understandings are part of the everyday lives of our children in order to prepare them for the world they will face in the years ahead.

What is it that drives us and how can we infuse sustainability thinking into our lives? Perhaps it is our everyday view of the world from space; or perhaps it is the speed by which we can bring what is happening in the world's hot spots into our living rooms in so many countries around the globe by computer, fax, or television. Whatever it is that drives us, on a local and global scale, we are moving toward better uses of these nodes of connectivity to improve the way we interact in our decision-making activities relative to the environment around us. Although our schools, colleges and universities still hold to classical disciplines such as life sciences, physical sciences, geological sciences, social sciences, still this spirit of connectivity is once more taking central stage as we evolve science for the new millennium. The National Science Education Standards, a national community-based document brought together by the National Research Council of the National Academy of Sciences, places front and center the importance of learning about science in context, of having the opportunity to learn science as scientists do it, through understanding inquiry, its role in scientific research, its promises to know about the natural world and its limitations. The recent Carnegie Report[4] on undergraduate education relates the need within the universities to be committed to new concepts of community and scholarship and deliver on the promises of the integration of the research and education missions within their walls. All of these reports bring us to new opportunities for synthesis of the building blocks, for building communities of learners and for ways to approach our learning to take into account the complexity of the systems that we study.

To the spheres of our blue planet, i.e., the biosphere, lithosphere, and atmosphere, we can now add the infosphere, coined by Berenfeld,[5] and described as a medium of information and communication of the developed knowledge about the natural and designed world that humans so freely interchange. No longer is this information only in the hands of the elite scholars, or the most powerful. Potential availability of information by way of the Internet for all peoples insures multiple ways to synthesize this information for understanding how to approach issues of sustainability in our world. It is possible these new areas of synthesis might be mining the most under-used capital resource, the human brain. Along with this new distribution of knowledge, we find new collaborations, formed among like-minded people into diverse communities of learners.

Where are the scientists? In the 1990s, scientists are taking their places within these communities to add their knowledge in decision-making processes. Collaborations are flourishing around tables to solve problems in schools, towns, state, national and international tables of leadership. More books are now being written by the scientific leadership that mirror a coming together of the traditional spheres of discipline knowledge so that words like collaboration, blurring the boundaries, interdisciplinary, institutes of change, consilience and yes, sustainability are now more common in print. Institutions are beginning to meet these needs by breaking similar barriers. This has led to new and important findings that are beginning to percolate through society.

Within governmental agencies, new themes of collaboration are addressing such issues as environment and global change, nature and the human society: the quest for a sustainable world (National Academy of

Sciences Conference), life and earth's environments (National Science Foundation), all themes that have arisen in the latter half of the 1990s. Universities and colleges have assembled interdisciplinary programs in environmental studies. Environmental courses are popping up that include not only the disciplines in physical, geological and biological sciences but also economics, ethics and religion. The first five years of the 1990s were spent writing the agenda reports for action; the more current moves are implementation of these concerns into actions that bode well for integrated policy relative to our concern about environmental sustainability.

Despite the often linear approach of the fields of scientific endeavor, scientists are developing new tools of visualization, modeling, nanotechnology, monitoring, database acquisition, and coherence that begin to formulate a systems approach to understanding the work in which we live. Translation of this strategic way to understand the interactions of various segments of society and look for the interactions that produce change, has brought local and global leadership to view their place in stewardship of our surrounding environment through new lenses. This newer view, a systems approach, allows us to model the system, discuss the intersection of beliefs and values, comprehend the concepts of change rates over time and perhaps predict our future with a solid input of the critical elements that will lead us to do the right thing. Decision-making is complex. Problems that relate to our work environment face us daily in our home and work environments. Just what is the science that we need to understand that will help us join these groups for change? What understandings will help us develop habits to sustain the quality and abundance of the natural world for the citizens of the future?

With this background in mind, there are some guiding principles that would be useful for a person to participate in sustainable thinking. Once in place, I would like to think the framework for a lifetime of learning is in front of every citizen.

- First is access to the knowledge derived from research in science disciplines. This is translated into conceptual content in the National Science Education Standards found on the National Academy of Sciences Home page (http://www.nas.edu). As citizens, we have opportunities to understand these concepts, as they relate to the environment, inquiry, and science in context, and we can make them available to every child.
- Second is the experience of doing inquiry, of actually participating in formal or informal education environments associated

with the way scientists do science. This insight of process leads to understanding and evaluating the promises and the limitations of science, demystifies how scientists do their work and connects each person with science as a way of knowing about the natural sciences so important to sustainability of the Earth system.

- Third is open access to all knowledge, but especially that on the so-called cutting edge of discovery. This is an ever-broadening area that involves books, museums, television, film, and other media, and the ever-expanding Internet. All are open ways for citizens to access information that is useful in the decision-making process that surround issues of sustainability.
- Fourth is to create environments and help facilitate all citizens toward actions and activities where we feel comfortable conversing about sustainability and the science behind it. These areas might be at local PTA meetings, the Rotary and other service clubs, college reunions, cocktail parties, church picnics, or in the comfort of the vast numbers of virtual chat rooms. These conversations must be characterized by respectful discourse, coupled with the art of listening, willingness to learn more and seeking the values associated with diversity of opinions.

At this point, we will have an informed citizenry that will have both the will and the way to create patterns for sustainability thinking in solving local and global problems. Let me be clear that science is only one of the elements within this complex system. However, it is central, it is organizing, and it is not so forgiving as we would like to think (for example, all would agree that extinction of species is forever!). It is quite realistic to think that some changes, based on our decisions, may well improve all of the other parts of the system, improve our options (taking into account our varying values), increase our economic well being, redistribute the wealth in sustainable ways, and pave paths of options for future generations.

It is not for us to totally define the science that one needs to be involved in reaching these conversations about sustainability; this too is an evolving field. It is important to recognize the need to listen to all kinds of input at the table and to recognize that scientists are still in a province of knowledge acquisition and distribution that is relatively primitive. Overarching themes however, help us organize the complexity of the systems involved in moving toward thinking about a one earth system.

SCIENCE FOR SUSTAINABILITY

Science is the organized systematic enterprise that gathers knowledge about the world and condenses the knowledge into testable laws and principles (Wilson, 1998). If we look at some areas of research that are currently underway, this informs us of the avenues scientists and the scientific enterprise are using to better serve the needs of humankind as we begin to uncover the needs for a more balanced approach to decision making policies at all levels. There are three overlapping areas that the National Science Foundation has recently identified under the thematic approach to research, "Life and Earth's Environments." The challenges to understanding the science in this theme often lie at the interfaces of where these areas meet. The areas are 1) ecosystem sustainability, 2) global change, and 3) human dimensions of environmental change. The science challenges inherent in these areas bring together the classical disciplines to a new degree of consilience. Concerns that have spurred these moves are 1) a lack of cohesiveness in our approaches to solving environmental problems despite the warning signals we are receiving, 2) concerns of availability of the basic sustainable earth system needs for the biotic populations such as safe food sources, clean and abundant water for consumption and agriculture, energy availability to fuel the expectations of technology, and cultivable soils for agriculture, and 3) the acknowledged increase in rates of change related to loss of biological species, human influence through industrialization activities on changing the world atmosphere, and accumulations of doing the wrong things that have overcome the built in resilience and buffering capacities of some areas of our major ecosystems.

Given these concerns and signposts of change, what is the science basic to the understandings that we should know? I believe that all citizens should have the opportunity to learn about the understandings in science basic to the fundamental concepts of sustainability. It is this sort of education that should reside in homes, schools and continue as a life-long process. Avoiding the science or thinking that science concepts are beyond an individual's ability to comprehend will lead to a false sense of leadership and may well lead to a belief that those same scientists, who bring forth these concerns, will develop miracle technological prosthetics as a way to replace anything really important in the natural systems that effect our lives. This is wrong! This will lead to indifference and placement of our children into a self-imposed ignorance, taking away their future rights, responsibilities, and freedoms to make informed decisions about an inherited "natural" world. Decisions we make now will decide whether that future inheritance is close to the natural sustainable world we now enjoy, or approach a propped-up, artificially grassed, desalinized system, with loudspeakers blaring out the sounds of brooks and bird songs!

SCIENCE PATHWAYS

The world's ecosystems are an intricate web of biological, physical, geological, and chemical processes that have slowly evolved over the three and one half billion years of the Earth's history. Humans have interacted with these ecosystems over a comparatively short period, but recent accounts (for example, Lubchenco, 1998) indicate that the human species has had impact on those ecosystems, reported in ways that are verifiable and consistent. Although many citizens have dabbled in understanding the complexity of these systems with amateur studies on, for example, birds, butterflies, stars, flowers, and frogs, it has been the scientists who have created domains of knowledge, through extensive and published studies, and developed the appreciation for the complexity of ecosystems through the exploration of the underlying processes common within the ecosystems. Although these understandings have resided in the classical domains of knowledge, there are new trends in research, such as the use of supercomputers for extensive modeling activities, satellites for observations beyond our wildest imaginations, genetic understandings that allow new ways to investigate ecosystem dynamics through a genetic marker toolbox, and new ways to understand the multiple stressors on these systems in taking too much out or putting too much in. We live in a time when humans are mobile, moving easily from ecosystem to ecosystem, unaware of being the conveyors of invasive species, from plants and animals to minute unknown pathogens, and these invaders are only now reaching numbers high enough to indicate their powers of creating change within the natural environment. One only has to hear the word El Nino to conjure up the amazing number of physical stressors that appear to result from atmospheric change. The scientists or observers of patterns of change can predict from models some of these changes but it does not change the enormity of the human distress caused by atmospheric disturbance. We continue to study the influence of human activity on these atmospheric measurements over time, with new visualizations of the earth from space, and studies of change over time. There are some bigger questions that are only beginning to be approached. What has been the carbon cycle over time, and how have humans influenced that cycle from the bottom of the oceans to the outer atmosphere?

One area of considerable interest is the concept of biodiversity—the bank of genetic diversity expressed in endless small and large forms of life or species, found in a multitude of associations called ecosystems. It is the multiple interactions between these species that has endlessly fascinated the scientists who continually look for large patterns of understanding. And it is the inherent need and rights of each of us to participate in the decisions that help us maintain the beauty of that complexity. Equally important are the small things that we hardly see, the forms that daily turn over the soils, breakdown the wastes, clean up after organisms die, recycle elements for larger plant and animal growth. Richness in biodiversity is beyond the big fuzzy bears and crocodiles; it lies in their surrounding ecosystem, in every backyard, in every shoreline tidepool, and deserves your attention. Concepts surrounding biodiversity are needed to understand sustainability for future generations. Every species will probably not be saved, but think how many we probably have saved once we understood the far-reaching effects of such pesticides as DDT? Eagles and hawks are high on that list. What we need are systematic strategies for managing and avoiding broad classes of environmental threats. As a species sharing this earth, we need to leave footprints of change that preserve the complexity, allow healthful coexistence, evolve thoughtful stewardship, establish a firm basis of education that conveys the science for change, and brings environmental talk to the level of coffee talk in every neighborhood!

EDUCATION PATHWAYS

Three major pathways are open for distribution and synthesis of knowledge toward developing and infusing processes and habits of minds to consider science and new relationships that interact with the Earth's environments. One is formal education in schools. The vast majority of our children can be reached in the K-12 environment and it is time to begin to develop this avenue. This country has a richness in the community college system and a further richness in the diversity of private and public four-year colleges and universities. Still fairly steeped in the discipline traditions, this venue however, holds more promise for innovation in meeting these challenges due to the keen shared sense of service and tendencies to "save the world" prevalent on many of the college campuses. The second major area is informal education within our vast network of museums and public media.

The third pathway is through thoughtful and innovative circumnavigation of the Internet. Already a buzzing source of information on sustainability, it must continue to be fertile ground for exchanges, for new ways to synthesize knowledge and for imaginative ways to bring systems thinking and sustainability together for the comprehension of world challenges. The Internet knows no national boundaries. The players are truly citizens of the World.

One overarching innovative pathway exists that I believe can act as a catalyst to attract numerous players into the World game. Scientists use well tested pathways in order to discover the underlying patterns in the natural world. With the connectivity of computers in the classrooms, it is now possible, with the help from scientists and teachers, to have classrooms full of students gather data, ask their own questions about their observations and participate in exploring/discovering their surrounding ecosystem. They might investigate data about birds migrating, butterflies landing, sea turtles laying eggs on the beach, frogs in ponds, and weather details at home and around the world. A suite of activities is rapidly developing, mostly for middle school and high school students, whereby students and scientists partner for the good of science and education. Teachers find their classrooms have a renewed sense of importance as data about various organisms, physical measurements of the ecosystems etc., are variously monitored, and the data sent to a scientific center for analysis and use. In most cases, the data for an enlarged area, including that school's data, is made available to numerous schools. This also may lead to school to school collaborations in sharing results and understanding world concepts.

REFERENCES

1. Clayton, Anthony M.H. and Nicholas J. Radcliffe 1996. Sustainability—A systems Approach. Westview Press, New York.
2. Lubchenco, Jane 1998. Entering the Century of the Environment: A New Social Contract. Science 279:491–497.
3. Wilson, Edward O. 1998. Consilience: The Unity of Knowledge. Alfred A. Knopf, New York.
4. Carnegie Report. 1998.
5. Berenfeld, Boris 1996. Linking students to the Infosphere, Technological Horizons in Education Journal 23:76–83.

CHAPTER 8

The Role of IUCN—the World Conservation Union—in Shaping Education for Sustainability

Frits Heselink and Wendy Goldstein

BACKGROUND OF IUCN

IUCN—the World Conservation Union, is an international conservation organization that has states, government agencies, and NGOs as members. IUCN was founded in 1948 as the International Union for the Preservation of Nature, IUPN. It became apparent that preservation that excluded people was not effective. So the organization adopted the name IUCN: International Union for Conservation of Nature and Natural Resources to reflect its new approach. Ever conscious of this mouth full of a title, IUCN added more recently a subtitle- "The World Conservation Union".

For IUCN "union" is the operative word. It is a Union of states, and environmental organizations; fee paying members, consisting of 913

Frits Heselink, Chairman, Commission on Education and Communication, IUCN, Gland, Switzerland. Wendy Goldstein, Head Environmental Education & Communication, IUCN, Gland, Switzerland.

members in 138 countries with 73 countries belonging as state members. The Union is also made up of a large volunteer network of some 9,000 individuals who contribute expertise and knowledge to IUCN's six Commissions.

The regionalized secretariat in 40 countries develops programs to serve the members. Programs vary from region to region, providing information, policy advice, and guidance to manage both conservation and equitable and sustainable use of natural resources. Countries are supported to develop environmental legislation, to prepare strategies for sustainability, establish a protected areas system, and to manage wetlands and forests for economic and conservation values.

Commissions also have programs resulting in species status reports such as the Endangered Plant and Animal Red Data Books, and scientific criteria for assessing the status of species, protected areas categories, international treaties, and training in environmental law and communication.

From the outset, IUCN played a role in educating about the environment, though its audience was mostly in the environmental field. IUCN never developed a high public profile like its off- shoot, WWF, the World Wide Fund for Nature. (WWF was originally set up to raise funds for IUCN).

THE COMMISSION ON EDUCATION AND COMMUNICATION CEC

The present Commission on Education and Communication has been through different name changes since it began its work in the late 1950s. The focus changed from only education to training and later to also embrace communication. The Commission's first main focus of activity was mostly European.

In 1970, IUCN with Commission input, developed the first internationally agreed definition of environmental education (EE), and contributed to the inter-governmental meetings defining EE in Stockholm and Tbilisi. The Commission tended to focus its efforts on policy advice, definitions, glossaries, formal education curricula, and activities for youth.

Now CEC is a regionalized global network with membership drawn from government, NGO, communication and education institutes, universities, and mass media organizations. Believing that there can be improved strategic use of communication and education as policy and project instruments, CEC provides advice and capacity building in this field. CEC therefore is oriented to the IUCN constituency—environmental government agencies and NGOs—rather than Education Ministries and formal education.

Environmental education and communication are disciplines based on rich research. Defining these terms is not where CEC focuses its energy. Instead, it is interested in the results of using the approaches and methods of these disciplines so that individuals, communities, and nations take responsibility for living sustainably. Getting there depends on how well we develop relationships to work together to solve the problems we face, and how well we can share what we know and discover on the way. Sustainability has a myriad of solutions created with people through a learning process. Once out of a formal education setting, the boundaries between education and communication become blurred, especially if you regard education as "to lead out of ignorance," and communication "to build relationships and shared understanding." Since the CEC niche is not in the formal education setting, we loosely use the term "education" or "communication" as umbrella terms for a wide variety of approaches to motivate engagement and enable people to reflect on how natural resources are used now and in the future.

BROADENING THE CONSERVATION FOCUS TO SUSTAINABILITY: AN INTERNATIONAL FRAMEWORK

IUCN has usually been on the cutting edge of conservation thinking. This derives from its rich network of experts, the experience of its institutional members and the on-the-ground experience of working in the field. At the international level, IUCN has worked on broad frameworks and policy advice. IUCN, for example, was one of the initiators of the treaty called the Convention on Biological Diversity.

IUCN, UNEP, & WWF recognized that conservation is not the opposite of development when they co-produced the World Conservation Strategy (WCS) in 1980. It emphasized that humanity, which exists as a part of nature has no future unless nature and natural resources are conserved. It asserted that conservation could not be achieved without development to alleviate poverty and misery of hundreds of millions of people. It emphasized that conservation includes both protection and rational use of natural resources, essential if people are to achieve a life of dignity and if the welfare of present and future generations is to be assured. Stressing the interdependence of conservation and development, the WCS gave currency to the term sustainable development. Sustainable development depends on caring for the earth, because the fertility and productivity of the planet are the basis of survival.

Ten years later, in 1991, IUCN, UNEP, & WWF published "Caring for the Earth—A strategy for Sustainable Living," based on a wide consultation. The Strategy was produced in the lead up to the United Nations Conference on Environment and Development, UNCED, in Rio de Janeiro in 1992, which produced Agenda 21 and the Convention on Biological Diversity.

"Caring for the Earth" is a strategy to underpin development that provides real improvements in the quality of human life and at the same time conserves the vitality and diversity of the earth. The goal is development that meets people's needs in a sustainable way. Most current development fails because it meets human needs incompletely and often destroys or degrades its resource base. We need development that is both human centered, concentrated on improving the human condition and conservation based, maintaining the variety and productivity of nature.

IUCN proposed a global strategy because we are all inter-linked. We depend on each other for maintaining the quality of our air, water, climate, oceans and species from which we derive so many life supporting products. To address poverty we need changes in our institutions, nationally and globally. Therefore we need to take mutually supportive actions towards a common goal, requiring cooperation and partnerships.

Defining sustainable development is difficult. "Caring for the Earth," suggests that the definition of sustainable development is ambiguous in the UN report, "Our Common Future" prepared under the guidance of Bruntland. For IUCN, sustainable development means, "improving the quality of human life while living within the carrying capacity of supporting ecosystems"—itself quite difficult to assess. The term "sustainable" has unfortunately gained currency and has been used in many inappropriate ways—like sustainable economic growth (appealing to governments and economists)—but nothing can grow indefinitely on a planet with finite resources.

In Caring for the Earth, a sustainable society lives by nine principles (Figure 1). These are not a prescription for action, but rather guide posts to be used on a journey towards sustainability. Our paths on this journey will be varied. The journey will definitely be a learning process, and we need to facilitate that learning. We know where we have made mistakes and can try to avoid repeating them, yet inevitably we learn by doing and will make mistakes. The principles and actions in "Caring for the Earth" are a basis for consideration and interpretation by different cultures and societies.

EDUCATING FOR SUSTAINABLE LIVING

Developing an ethic and changing attitudes and practices are some of the Strategy's means to sustainability. To change attitudes and practices Caring for the Earth proposes to:

- include in national strategies for sustainability action to motivate, educate and equip individuals to lead sustainable lives

Principles for Sustainable Living

Founding principle providing the ethical base for the others

1. Respect and care for the community of life	Duty to care for other people and other forms of life now and in the future.
Criteria to be met in achieving the ethical principle	
2. Improve the quality of human life	Enable people to realize their potential, lead lives of dignity and fulfillment.
3. Conserve the earth's vitality and diversity	Conserve life support systems, biodiversity, ensure uses of renewable resources are sustainable.
4. Minimize depletion of non renewable resources	Use less, re-use, recycle, switch to renewable where possible.
5. Keep within earth's carrying capacity	Including human population and level of consumption.
Directions to be taken in working towards a sustainable society	
6. Change personal attitudes and practices	Education formal, non formal and training to provide technical capacity.
7. Enable communities to care for their own environments	Communities mandated, informed and empowered to care for their own environment and not to degrade that of others.
8. Provide a national framework for integrating development & conservation	Consistent policies, institutions, and information to advance in a rational way.
9. Create a global alliance	No nation is self sufficient, an alliance is required for global issues, and to share more equitably.

It is recommended that new strategies for sustainable living based on the nine principles are developed and national development plans are reviewed.

Figure 8.1

through integrated and consistent campaigns, non formal and formal education;

- make environmental education (including social education) an integral part of formal education at all levels;
- determine the training needs for a sustainable society and plan to meet them;
- development assistance is urged to give more support to environmental education and training. It is suggested UNESCO,

> UNEP, and IUCN establish an international clearinghouse for environmental education.

However, the Strategy recognizes that people act in unsustainable ways for many reasons other than lack of knowledge. There is a hierarchy of needs that must be met before sustainable use may be possible and certainly before an environmental ethic can be applied. People must have security first.

While in many economically poorer countries, poverty causes environmental destruction, in turn that destruction undermines hope for the future. People trapped in this cycle know perfectly well that the environment is being degraded. Education to help them may need to be directed in the main to other Parties, as illustrated in the Dutch example (see Box 1).

In more affluent groups and countries, unsustainable practices occur because of ignorance, greed, lack of concern, lack of policies, incentives for wastefulness and structures to help people. Even those who accept the need to live differently often fail to follow their ideals.

The Dutch contribute, in global terms, a high proportion of GNP to development assistance (approximately 1%) and to environmental aid. Yet in this simplified example, from the Netherlands we can see the complexity and impact of the economically developed countries on those less economically developed, and why we have to deal with equity issues to achieve progress. Dutch beef is raised on crops grown in SE Asia and Brazil. In many cases the production of these cash crops led to clearing of forest or rainforest. The beef, produced with European and Netherlands agricultural subsidies, could be exported to Western Africa and sold cheaper than locally grown beef. As the price was low, it was not worth it for a local farmer to kill his cattle for market. So the cattle were kept and the herds grew causing over grazing and contributing to threats of desertification. (Meanwhile Dutch aid is directed at stemming desertification and saving rainforests!) Communication, education, and advocacy are required amongst Europeans—decision makers, farmers etc—to change subsidies and policies to stop the beef dumping and so assist the West African farmers. Indeed the subsidies that permitted this beef dumping have been reviewed.

This example arises from a study done by the Netherlands National Committee for IUCN called "The Netherlands Impact on the World Ecology." The intent is not to single out the Dutch but if other nations had so examined their impact on the world ecology it would have been possible to draw from any developed nation).

Caring for the Earth recognizes that education needs to go hand in hand with other measures, changes in instruments, structures and institutions. This means policies may have to be aligned or changed to be consistent, for example in Australia there is a subsidy to clear land for agriculture, as well as scheme to plant a billion trees. Facilities need to be made available that make it easier for people to change practice, like having regular public transport or waste recycling capacity and drop off points.

The Strategy also recognizes that it is necessary to engage and enable communities to participate. However to enable communities to care for their own environment there are often many administrative and institutional changes required. People may have to be given rights to land tenure, or even to participate. These changes are an important part of creating conditions for people to gain more benefits from the management and use of natural resources. These institutional changes need to go hand in hand with other appropriate support and effective communication and education. Communication and education is the ingredient to motivate, understand perceptions, build relationships, identify shared goals, to work out solutions and share knowledge and understanding. The inter-relationship between these many factors is shown in Fig. 8.2 in an example of an IUCN project in the coastal area of Guinea Bissau, West Africa.

On the river Buba in Guinea Bissau, poor communities gained little benefit from the fish in the river. Furthermore, it was likely that fish stocks would decline due to over fishing by fishermen from neighboring countries, who in contrast to the locals, had nets, storage facilities, and powerful boats. Some of these foreign fishermen cut the local forests to smoke the fish before taking their catch home. This activity was reducing forest cover around the river and coast line. As the local community received only limited benefits from the fish resources there was little incentive and capacity to manage them or protect the fish from foreign fishermen.

Educating the communities to empower them to manage the fish had to go hand in hand with enabling them to gain increased benefits from the fish. When practices for sustainable use of the fish were adopted there was a win/win situation for the villagers and conservation.

A sketch of the inter-relationships of the communication & education with technical, institutional, financial and legal inputs is shown in the simplified exposition in the box. Communication is used to build relationships with many groups and organizations so as to solve the problems with the community. The local population was motivated to enter into co-management of the river resources when the benefits to them became clear. The result has been more economic benefit from less fish, and a strong community involvement by the women in economic enterprise. The women wanted to learn to read and to help other communities develop like them.

At the time of writing this, Guinea Bissau is undergoing civil strife and the fate of the Buba River people is unknown.

In what ways has the IUCN Commission on Education and Communication (CEC) taken up the challenge of educating for sustainability?

Problem	Communication Education/training	Financial/Technical	Institutional/legal
Fish populations dropping in Buba River due to over fishing by foreign fishermen	Build relations and trust with communities along both sides of the river to interest them in better management of the river resources. Listen to communities perception of the problem and their needs	Identify technical and financial inputs	
Local GB people cannot catch enough fish themselves because of a lack of nets, boat motors	Develop shared goal Develop consensus on how to manage micro credit system themselves. Develop skills in financial management	Provide 1 net for each community as a way of showing good faith and building trust to engage in the project Provide funds for micro credit—boat motors etc	Develop village managed mechanism for storing and handling cash loans
Assess sustainable levels of fishing	Motivate national research institute to develop management plan for sustainable fishing in cooperaton with community. Engage fishermen in sharing knowledge on the breeding cycle of fish and to give advice on the fish management plan. Education on why the management plan is needed		National research institute undertakes research Community information listened too building trust with this agency
Community needs to own management plan, and have regulations to control foreign fishermen Implement management plan	Consensus on plan. Communicate results that during the breeding period no fishing is to take place Communities and foreign fishermen informed of the regulation	Fish management plan agreed by government and communities	Government makes a regulation for the Buba River for a no fishing period Government supports enforcement of the ban as do locals

Problem	Communication Education/training	Financial/Technical	Institutional/legal
Improve market value of fish and preserve the catch	Contact and interest women in processing fish Engage NGO in developing salting & packing techniques Teaching salting/drying process and packaging Revival of skills in peppering fish Develop skills In gardening for local vegetables and peppers.	Import salt Mali Packaging techniques Value of fish product increased 7 fold Village based salting, packing and storage shed is built by community with small finanncial input	
No outlet for products for fish & vegetables	Engage local authority to approve and supervise market Engage NGO to promote market on radio program in capital city- 200 kms away-	Financial input by community	Local government to set up community market NGO partnership for radio
Handling marketing	Engage Ministry of Education to teach numeracy and literacy	Classroom developed	Adult literacy and numeracy classes
Forest degraded by fisherman smoking fish	Engaging NGO to design a new technology for smoking fish. Informing communities and foreign fisherman	Design of more efficient smoking method. Financial support to introduce the stove	NGO partnership
Forest degraded by need to replace dug out canoes every 3 years	Engage NGO to develop plank boat building Teach new skills boat building	New technology developed for boat building Financial support	NGO partnership

Figure 8.2

PROMOTING COMMUNICATION AND EDUCATION AS INTEGRATED POLICY INSTRUMENTS

As the Buba river case illustrates, communication and education are most effective if they are used hand in hand with other inputs, rather than as isolated activities.

The Commission on Education and Communication, CEC, advocates for a more strategic use of communication instruments (which embrace education, instruction, marketing, public relations, information) and seeks to provide services and capacity building in this area.

While CEC definitely supports reform of formal education to deal with sustainability, our niche has been to argue for the use of communication and education as instruments for policy.

If governments are going to achieve policies for sustainability, communication has to be an integral part of all policy or program phases. Communication plays different roles for government and NGO in the policy cycle, from the outset when putting an issue on the agenda, during policy formulation, during implementation and later when monitoring the policy.

During the signaling or agenda setting stage, NGOs or community groups are actively advocating the importance of an issue. In this phase governments are listening, using communication to assess community interest in the issue through polls, interviews with stakeholders and by reviewing how much attention the issue has in the mass media. When governments see the issue is important, they move to form policy.

In the policy formulation phase governments use communication to consult or interest stakeholders to participate. Stakeholder participation helps governments to develop more realistic and socially acceptable policies and to define the most effective combination of instruments. For example, in Australia a hazardous waste management policy was negotiated over 18 months with key stakeholders. Negotiations worked towards a win/win situation. While the process seemed time consuming and expensive, the policy has backing from key community sectors, industry and environment alike giving it a greater chance of effective implementation.

When it comes to implementing policies, governments need to use communication and educational approaches in a strategic mix with financial incentives, structural changes, and regulations according to the level of people's voluntary intent to change.

The idea is that the community ultimately takes responsibility for the policy. Communication plays a vital role to build relations with partners and stakeholders. It is these groups who actually implement the educational activities with their target groups. Governments also use communication to support and expand the networks active in the issue and to mobilize knowledge and learning to help the groups in their work. Governments can also lever involvement in implementing policy by providing financial support to NGOs and community groups and by funding community facilitators.

PROMOTING A FOCUS ON KEY GROUPS WHO CAN INFLUENCE THE ISSUE

Environmental education has been evolving since 1970. In Fig. 8.3 the trends in substance, approach and target audiences are shown. From a focus on schools and children we are increasingly moving to include other sectors of society in educational efforts. From an orientation to "nature study EE," studying in and about the natural environment, attention is shifting to encompass more development aspects and taking an economic & social viewpoint in relation to natural resource use. The emphasis CEC is trying to make is in the region on the lower right hand side.

From the 1970s environmental education concentrated mostly on nature study and developing a love of nature. Work focused more on primary education where more integrated approaches were possible. Later, issues like energy use, pollution and human use of the environment were introduced, including urban environmental issues. Development was presented as destroying nature, with little recognition of how to strike a balance. Integration of the environmental let alone sustainable development agenda in a holistic way was constrained by the disciplinary and institutional formal education system, particularly at secondary and tertiary levels. There were arguments about whether EE should be integrated in all subjects or taught as a separate subject.

Now, as we move to education for sustainability, considering social and economic factors are as important as the ecological. Educational approaches have concentrated largely on the formal education sector in the past, yet adults are creating the havoc right now. Increasingly we have to turn our attention to adults in the boardroom, factory floor, international and small corporations, farm, unions, consumer groups and local communities to bring about sustainability. Many other groups in society such as businesses, NGOs, consumer groups and so forth are taking up education for sustainability.

CEC's contribution is to encourage more focus on key groups who can make a difference to an issue in a role as either "educators" do or as "learners" do. To engage adults we have to be aware of their ways of learning and how to be able to attract them in their busy lives. Not only do we

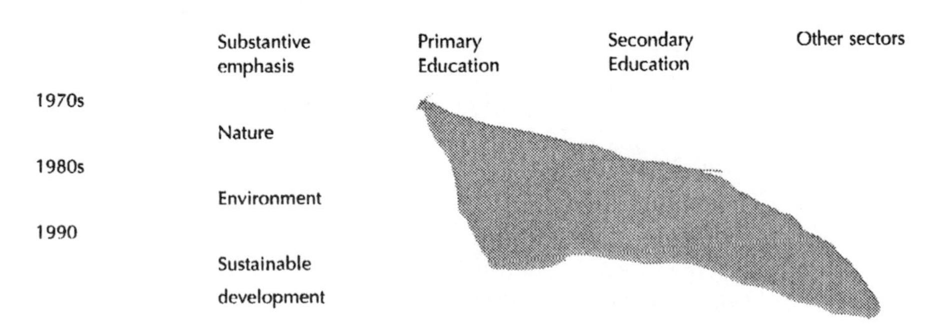

Figure 8.3

need to use marketing approaches, and have a credible story, but also we have to be less idealistic about the power of rational argument alone. Our approach will be more how do we reach the result wanted? We will appeal to emotions and personal and economic benefits. Being conscious of cynicism amongst adults it will be important to show that others are involved and making changes too. Systems will be in place to make whatever new actions are decided on, easier to do.

RAISE THE ISSUE OF EDUCATING FOR SUSTAINABILITY HIGHER ON THE GOVERNMENT'S AGENDA

For Agenda 21 governments have made an agreement, but not law. What they do depends on leverage and political will. The United Nations Commission on Sustainable Development, CSD, is a mechanism to lever action and is a forum where governments can learn from each other.

Various United Nations agencies have responsibility to be task managers for the many chapters of Agenda 21 and to report to the CSD on progress. There was little dissension for chapter 36 on Public Awareness, Education and Training at Rio. Indeed the words "awareness," "education," and "training" occur so numerously throughout the Agenda 21 that they are mentioned second only to "governments." Yet a report of the Secretary General of UNESCO as Task Manager for chapter 36, in 1996 stated that education, awareness and training risked being "the forgotten priorities of Rio."

To work against the threat of education being "the forgotten priority of Rio," the IUCN Commission on Education and Communication has played a role in different ways to try to increase attention to this issue.

IUCN, UNESCO, and UNEP held a series of regional meetings to review progress on strategies to use communication as a policy instrument and develop national action plans for education for sustainability in Europe (1993), Latin America (1994), and Asia (1995). These meetings revealed a concentration of effort on the formal education sector and weak integration of education and communication as a policy instrument. By focusing on the school sector educational efforts are usually separate from rather than being an integral part of solving problems related to a policy (or of a project). Almost universally, there was weak political support and therefore resources for communication and education. National EE action plans, where developed, languished without commitment to implement them, at times suffering from changes in political parties and in their interests. While some countries made financial commitments, it was more generally true that funding was often preemptory and inconsistent.

To put chapter 36 higher on the political agenda of governments, CEC advocated for a CSD work program, which was adopted. It proposed that this "international work program on education for sustainable development raise the status of education and communication on the political agenda, increase understanding of its basis and framework, support national actions plans and work to increase the amount of funds allocated to support education and communication." The work program, "should make clear provisions concerning the periodic assessment of progress and impact." To this end, CEC prepared its own position paper that it circulated to IUCN state government members and to CEC members to disseminate to their own government delegations. Some CEC members were able to influence their government's positions. The European Union statement, influenced in advance, in 1996 was the most supportive and called for a CSD work program.

CEC suggested that CSD might call for "an alliance of UNESCO, UNEP, the World Bank, UNICEF, UNDP, WHO, and international NGOs such as WWF and IUCN and others to develop and implement such a work program." It argued for the World Bank and Bretton Woods Institutes to review their investments in education to better support education for sustainability. The latter measure was in large measure agreed to by CSD in 1996.

One element of the CSD work program is to define the concepts for education for a sustainable future. To this end, UNESCO and the Greek government held an international conference in Thessaloniki in December 1997. United Nations agencies and others such as OECD, the European Union and IUCN CEC contributed input to a background paper.

CEC worked on providing input to key issues defined by UNESCO for the conference. This was derived from an Email request for suggestions from CEC members. A draft paper was compiled. CEC then went on to develop this position and recommendations for further discussion at Thessaloniki. At the conference an international group of CEC members worked to develop recommendations for the CSD work program.

CEC has also been a sponsor of an Internet debate on education for sustainable development that has engaged experts from around the world to share their thinking on what education for sustainable development is about. CEC will publish the results of the discussion in 2000 and the publication will be available from IUCN.

CSD has dealt with chapter 36 each year in reviewing progress on the work plan, so it gets attention. However, progress on the work program is slow and as yet there is not really a concrete plan of action. Apart from preparing input to the CSD, there appears to be a very limited process of cooperation amongst the international agencies. Efforts to find synergy and a strategy of working together are not apparent and CEC has backed off making much input into this mechanism.

The World Bank has reviewed its investments in education, some 2 billion dollars, and has recognized it doesn't use them to stimulate education for sustainability. How do we lever those funds to this work when governments as the clients of the Bank are responsible for determining what they do with their loans?

In other forums, CEC has attempted to raise the profile of education for sustainability, such as at the American Summit on Sustainable Development. CEC members used the position paper in advocacy to their governments. The recommendations from the Summit were picked up in advocacy to the Parliamentary Conference of the Americas. The CEC North American regional chair and his organization, Learning for a Sustainable Future, LSF, Canada and CEC produced a "Legislative Framework on Education and Communication for Sustainable Development: A model." We did not evaluate results from this intervention and hope that the material made available has been useful. In retrospect we should have made the material more generally available to the CSD and education networks.

BIODIVERSITY EDUCATION & COMMUNICATION POLICY ADVICE

The main approach for IUCN into sustainable development is through biodiversity or conserving the earth's vitality and diversity.

IUCN therefore plays a role in giving policy advice to governments in their negotiations on actions to take under the Convention on Biodiversity, the most comprehensive treaty dealing with life on earth. Ratified by over 140 countries it contains a far-reaching agenda for economic, environmental and social changes across the globe. Because governments ratify the Convention in their own parliaments it becomes law, unlike Agenda 21.

The Convention on Biological Diversity is based on three principles:

- the conservation of biological diversity;
- the sustainable use of biological resources;
- the equitable sharing of benefits from use of biological resources.

The Convention takes account of indigenous knowledge and rights. While respecting sovereign rights to biodiversity, the treaty obligates Parties to an international duty and responsibility to conserve their national biodiversity.

Public Education and Awareness, article 13, of the Convention states that the Contracting Parties shall:

13a) "Promote and encourage understanding of the importance of, and the measures required for, the conservation of biological diversity, as well as its propagation through media, and the inclusion of these topics in educational programs; and

13b) Cooperate, as appropriate, with other States and international organizations in developing educational and public awareness programs, with respect to conservation and sustainable use of biological diversity."

There are no escape clauses in 13a), so it is assumed that in agreeing to the wording of the Convention, governments saw this as a vital activity to undertake. Therefore governments have a duty to encourage understanding in the public.

To influence and guide decisions on Article 13 by the Parties, CEC organized the following actions and events: (see appendix).

1. Recommendations to SBSTTA 1997

An international workshop at the Global Biodiversity Forum preceding the meeting of the technical body for the Convention, SBSTTA (Subsidiary Body on Scientific, Technical and Technological Advice) 1997 drew up recommendations to SBSTTA.

Key workshop recommendations requested the Parties to regard communication as an instrument of policy of equal standing to that of legal and economic instruments, and argued for the professional skills of communicators and educators to be included in the early phases of the policy process. As biodiversity conservation is a social issue, the recommendations urged SBSTTA to include professionals in the social sciences and communication in its technical deliberations.

Drawing from the workshop lessons on effective ways to engage people in biodiversity issues, practical advice was prepared for the Parties. For example, the Parties were reminded that education needs to be based on traditional and people's own knowledge and perceptions. This was to correct the erroneous assumption that simply presenting biodiversity scientific information can bring about action. Attitudes based on values and beliefs, social relationships and structures play significant parts in determining the course of social action that adults deem most appropriate in their particular circumstances (Ham and Kelsey).

While Article 13 suggests propagating biodiversity through the mass media, a significant body of educational research disputes this approach in bringing about broad social change. "Rather than attempting to reach vast generalized audiences it emphasizes the importance of designing educational initiatives for specific groups within specific contexts. This targeted approach—which has much in common with the concept of market segmentation used in the corporate sector—is supported by contemporary models of learning which argue that knowledge is dependent upon context and actively constructed and reconstructed within the world of real practice." (Ham and Kelsey, 1998)

The above quote comes from a study in Canada on the marriage of educational theory and practice in biodiversity education. A CEC member triggered this study after being in the CEC workshop.

2. CEC Recommendations to COP

Building on the recommendations to SBSTTA that arose from the Global Biodiversity Forum, CEC developed recommendations for the next Conference of the Parties COP. An E-mailed request yielded a high return of responses from CEC members; 50 out of 300 from the international network. A CEC member drew together the first draft from these recommendations, which was circulated to IUCN members. A copy of the recommendations was sent to the Biodiversity Convention Secretariat. The background paper circulated to the Parties UNEP/COP/4/19 had a high correlation with the CEC recommendations.

3. CEC Members Lobby Their Country Delegations

CEC members were also e-mailed the recommendations and asked to gain support for them from their government delegations. Known examples included Brazil, Spain, Netherlands, Argentina, Ecuador, Peru, and Thailand. In some cases it was possible to give additional advice on draft government position statements, demonstrating the effectiveness of CEC to reach from grassroots to government.

Lessons and recommendations—Workshop on "Public Education and Awareness: Putting it into practice" at the Global Biodiversity Forum preceding the fourth Conference of the Parties COP.

A rich array of expertise and analysis at the workshop drew out a high degree of accord on basic principles for educating. The learning from these cases was made available at the COP and on the Internet.

The accord on managing learning specifically revolved around:

- Defining the issue: making biodiversity realistic and tangible;
- Defining who should be involved;
- Understanding perceptions—motivation;
- Defining a shared goal among stakeholders—the result;
- Building on local knowledge and culture;

- Finding ways to involve stakeholders to plan, implement and evaluate policy and programs;
- Facilitating networks at all levels.

Recommendations to the parties were draw out of the workshop and emphasized the strategic role of communication, the need to allocate resources, and to build capacity to manage communication and education.

4. Advocacy at the COP

An informal meeting of government delegates during the Conference of the Parties, COP was used to focus on CEC views on Article 13 and to bring the findings of the Forum to the attention of the Parties.

5. Capacity to Manage Communication for Biodiversity

Translating international negotiations to a practical level for governments and NGOs will continue to be part of the CEC program. In South America, Argentina, Peru and Colombia IUCN & CEC are providing assistance to use and plan communication as part of national biodiversity action plans.

CEC has provided communication management training to national biodiversity action plan managers in South America.

In Europe, CEC works as one of the program coordinators for the Pan European Biological and Landscape Diversity Strategy theme on public education and awareness. In particular it has developed a professional training program for government and NGO communicators/educators.

Providing training for IUCN members and improved access to the expertise in biodiversity education and communication will be a continuing focus of CEC's work. This will include mobilizing knowledge and expertise as well as well as mentoring and training.

CONCLUSION

Educators often criticize IUCN's instrumental view of education. The word instrument suggests manipulation perhaps. We also emphasize communication and this is often interpreted as synonymous with campaigns, mass media and telling people what to do.

CEC argues for being realistic. Learning and change depend on context and emotions, not just on scientific information about biodiversity or teaching skills. People will conform to what is happening around them, and of being accepted.

Communication is about seeking shared understanding between the way you see the situation and the way I see it. As educators we facilitate expanding the area of common meaning. What inputs and challenges can educators make to cause people to reflect and consider their actions, thoughts and attitudes and that of their social context? What inputs can we make to help people learn in a process of planning, action and evaluating?

Policies do not work well when elite groups decide, announce and defend them in top down communication, a practice more common than not. Buy in or ownership happens when people are engaged to think with you and be active. To engage people in stakeholder negotiations in order to develop shared goals or policies requires effective communication and management of the learning process.

By understanding possibilities for change, communication can be associated with appropriate support and incentives. Not all people will change because they are educated. Above all, we need to recognize that needs must be met, and education has to work in that reality.

REFERENCES

IUCN UNEP, WWF (1992) Caring for the Earth A Strategy for Sustainable Living, IUCN Switzerland.

The Netherlands Impact on the World Ecology, IUCN Netherlands National Committee.

Report of the Secretary-General to the Commission on Sustainable Development (1996), (1998) Chapter 36 of Agenda 21: Education, Public Awareness and Training.

UN Economic and Social Council, Review of cross sectoral clusters: Education, Science and the Transfer of Environmentally Sound Technology, with particular reference to Agenda 21, chapters 34, 36, and 37. Promoting education, public awareness and training (chapter 36 of Agenda 21) E/CN.17/1997/L.9. Commission on Sustainable Development Fourth Session 18April–3 May 1995.

IUCN CEC (1997) Educating for Sustainable Living, Imagine tomorrow's world. Prepared for the international conference on Environment and Society: Education and Public Awareness for Sustainability, organized by UNESCO and the Greek government December 1997.

IUCN Commission on Education and Communication CEC (1996) Statement on Agenda 21 Chapter 36 to the Commission on Sustainable Development, IUCN paper.

IUCN Commission on Education and Communication (1997) CEC's contribution to the CSD work program.

IUCN CEC (1997) Education and Communication: the forgotten priority of Rio? IUCN Paper.

IUCN (2000 in press) Education for Sustainable Development Debate, CEC IUCN.

LSF & IUCN CEC (1997) Legislative Framework on Education and Communication for Sustainable Development: A model.

Ham, L. and Kelsey, E. (1998) Learning about Biodiversity: A first look at the theory and practice of biodiversity education, awareness and training in Canada. Working Document The Biodiversity Convention Office, Environment Canada.

APPENDIX

Summary of CEC action towards influencing the COP4 decisions on Article 13 Public Education & Awareness

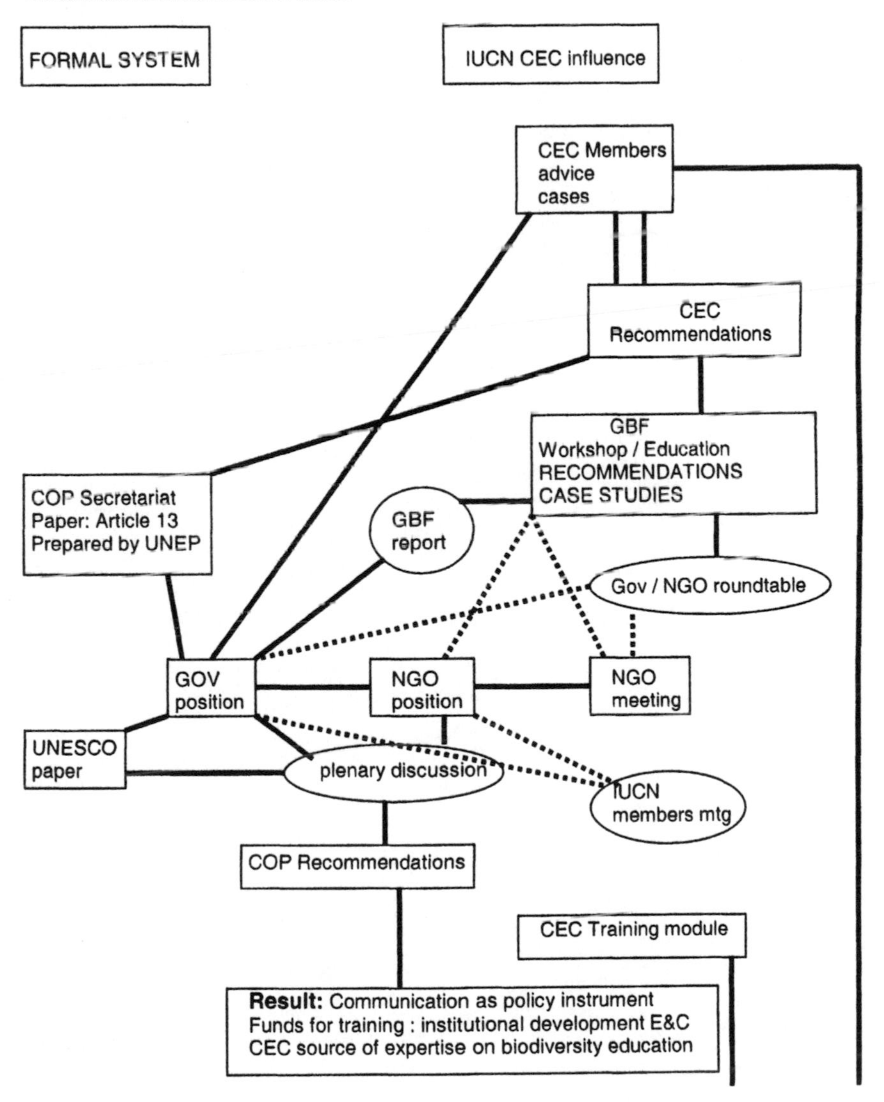

CEC on ground support for biodiversity communication and education strategies : training, case studies = improved implementation

CHAPTER 9

Exploring Sustainable Communities

Mary Paden

"Ask students to imagine what kind of community they would like to live in as an adult. Explain that the object is to collect as many ideas as possible—nothing is too small, too big, or too crazy for consideration. Tell students that this technique has been used in real-life cities with great success, as they will learn later."

So begins a lesson in "Exploring Sustainable Communities," the 12th secondary school teacher's guide unit in the *Teacher's Guides to World Resources* series published by the World Resources Institute. The guides are based largely on policy research done at the Institute, but this guide was also inspired by the three-year effort of the President's Council on Sustainable Development (PCSD) to create a set of recommendations to promote sustainability in America. The Council, with its "balance" of environmental, business, labor, and government interests, seemed like an unwieldy group unlikely to agree on a common set of recommendations.

What they were able to agree on—after many weekend retreats and marked up drafts—was a common vision. "Our vision is of a life-sustaining Earth," they wrote. "We are committed to the achievement of a dignified, peaceful, and equitable existence. A sustainable United States will have a

Mary Paden, (At time of writing the author was) Environmental Education Director, World Resources Institute, Washington, D.C.

growing economy that produces equitable opportunities for satisfying livelihoods and a safe, healthy, high quality of life for current and future generations. Our nation will protect its environment, its natural resource base, and the functions and viability of natural systems on which all life depends."

Once that vision was crafted, PCSD members were able to start productive discussions of how to achieve it. Disagreements continued of course, but there was at least a pull, something to work towards.

The PCSD was greatly inspired by a meeting held in Chattanooga Tennessee, which had undergone a city-wide visioning process 20 years earlier. In the late 1960s, Chattanooga had the distinction of being named America's most polluted city. In a remarkable turn-around, Chattanooga is now a showplace of a sustainable city. The change began with a city-wide visioning process that involved neighborhoods, schools, city officials, and churches, gathering in newsprint-walled meeting rooms and discussing what they would like their city to be. Ideas were all recorded and later grouped and sifted, regrouped and discussed further. Some common themes emerged. The city would be clean, safe and friendly. There would be a walking park along the river and a new freshwater aquarium would attract tourists. Lower income residents would have access to rehabilitated and new homes. Clean industries would provide jobs and taxes. Frequent electric buses would transport workers and shoppers along the long narrow downtown area.

This vision, a dream 20 years ago, is now a reality after many years of hard work. Recently Chattanooga held another round of visioning meetings called Revisioning 2000. The 2,559 new ideas generated have been boiled down to 300.

This type of visioning exercise is happening in cities across the country. Our Teacher's Guide, "Exploring Sustainable Communities," attempts to bring this process into schools, where it has rarely been done.

Most students have been asked what they want to be when they grow up. But very few have been asked what they would like their community to be like. And most students initially have a hard time addressing the question.

The first obstacle to envisioning a desired future for the community is understanding that change happens. When your whole life is only 15 years it can be hard to understand change. The first exercise in "Exploring Sustainable Communities" has students investigate changes in their communities over the past 50 years—physical, social, economic, and environmental changes. They are asked to interview parents and elders and collect photos and newspaper clippings. Many American communities have changed extensively over the past 50 years: new jobs have attracted workers, suburbs have mushroomed, inner-cities have crumbled, some have revived, waves of different ethnic groups have moved through neighborhoods, woods have been cut down, crime has risen or fallen. Many of the changes have reflected change in the global economy, rising environmental consciousness, and changing technologies.

But why does change happen? Is it just the random sweep of history or do people make it happen? This question brings up the second obstacle to envisioning a sustainable community: the notion that we cannot influence our environment. Public opinion polls show that Americans have a very low level of confidence that government—even local government—will act to improve their quality of life. However, Americans have a strong sense that they can improve their lives through hard work and a little luck. We believe individuals and families can grow and prosper, but that is as far as our sphere of influence goes. To think that we can shape our *communities* is a stretch for most adults, as well as students.

Yet historically many communities had founders or early leaders who purposely set them on a path of attracting certain types of industry, or philanthropists who encouraged certain arts for which the community became known. They set the tone and character of the community according to their own character and beliefs. Current civic leaders and elected officials may opt for business as usual or conduct their business poorly, but often enough they make a mark that changes the direction of the community.

Once we agree that change happens and that it can be directed by people, we often hit the third obstacle: why bother with a vision? Why not just solve the problems? Problem solving is indeed an important skill taught in many classes. Students who learn it well can look forward to easier and more creative lives at work and at home. Isn't it enough to teach them to solve problems?

Visioning doesn't replace problem solving. It is a powerful tool with another use.

Visioning generates a common goal toward which a group can move. It offers a possibility of fundamental change. It generates creative thinking and sometimes even passion. By agreeing on a vision that is set in the future, participants can often set aside current differences in objectives or approaches and find an area of agreement (of course, differences may arise again in the discussions about how to achieve the vision). In moving towards a vision, many problems will require solving.

In our classroom community visions, we first ask students, "What would your community be like if you had the power to make it any way you wanted it?" We try for specifics. Where would people live? Where would they work? How would they get to their schools and workplaces? What would they do for fun? What kinds of houses would they have? Who would their neighbors be? What kinds of jobs would they have? Students are asked to record their vision—whatever it may be—and the teacher sets it aside for a few classes to be revisited.

The teachers guide their study through several lessons that examine urbanization trends worldwide and some of the typical urban problems in both fast growing cities in developing countries and transitioning rust-belt cities in the United States. They also role play city administrators or organized neighborhood groups to solve a set of case-study problems of urban transportation, housing, water pollution and air pollution. And they hear a few ideas about sustainability. Finally, they revisit and revise their vision.

What kind of communities do kids want? Many of the visions, of course, contain provisions for additional teenage liberties such as a comment from Rockville, Maryland that "middle school will start at 9 AM because adolescents need their sleep." Students at some schools were quite tough on crime and heavily into rules for communities (probably reflecting the rule structure in their own lives at home and school). In general the community visions call for personal safety, open green space, friendly neighbors, places to walk and bike, public facilities such as libraries and museums, places for young people to gather, and easy transportation. Their communities of the future will have good jobs and time for family and fun. Stresses are reduced. People are helpful. Here are a few of the specifics posted on our web site that were sent in by teachers who used the guide in class:

The first Community Vision from Anoka, Minnesota is an example of a polished statement that went through a process of review and revision. It was sent to us by one of its participants, Bill Mittlefehldt, a teacher at Anoka High School, who also participated in the PCSD working group on education.

"Anoka is a river town in the metropolitan Twin Cities of Minnesota. The Mayor, Peter Beberg, and the Anoka City Council have worked together over the last ten years to involve more students and citizens in their community. Anoka High School, the largest Minnesota school, now gives credit to students who work on a number of these projects, task forces and committees. Anoka's leaders in the city, school district, business, health and environmental groups have been working with the community and schools to build community partnerships aimed at Anoka's Community Vision."

"Anoka's vision grew from perceived problems in the late 1980's: juvenile crime, cruising, poverty, health, economic development, and environmental quality. The city council took the recommendation of the Anoka Economic Development Commission and started a strategic planning process to prepare Anoka for a more sustainable and dynamic future. In 1990, Anoka's leaders teamed up with the University of Minnesota's Project Future to involve 4,000 students and citizens in visioning processes for their community. The result of this assessment was Anoka's Community Vision Statement."

A COMMUNITY VISION FOR ANOKA IN THE YEAR 2010: A UNIQUE COMMUNITY

Anokans recognize and appreciate the distinctive character of their community and take pride in the neighborhoods, families, and enduring friendships. In 2010, Anoka will be a community which:

- Has a unique downtown that preserves the "Main Street" image, blending merchants, professionals, residents, and visitors in an active trade and service center;
- Recognizes its heritage, and plans for a future that preserves and celebrates its history, culture, and distinctive physical features;
- Is clean and takes pride in the appearance and condition of buildings, parks, streets, and natural features.

IMPORTANCE OF HUMAN RESOURCES

Anokans recognize the importance of their human resources, the character provided by each generation, and the power of human networks to provide for the needs of all citizens. In 2010, Anoka will be a community which:

- Encourages cooperation with and support for the cities that make up the "Community of Anoka," combining energy and resources when possible to create the best situation for all cities involved;
- Encourages a friendly, neighborly, and safe place for children, teens, adults, and senior citizens to live;
- Deters drugs, crime, and violence, with a Police Department working together with citizens to maintain a peaceful and secure environment;
- Values and encourages "volunteerism," enabling citizens to enjoy the rewards and see the results of giving their energies to the improvement of the community;
- Responds to the needs of youth with a variety of entertainment options, personal development opportunities, and social activities;

- Views persons of diverse ethnic and cultural backgrounds, physical and mental abilities, age, and economic condition as valuable members of the community;
- Recognizes the importance of leisure time and fun for well-balanced living by promoting diverse and exciting entertainment and activities for all ages.

ECONOMY AND OPPORTUNITY

Anokans recognize the importance of an economy which provides opportunities for all residents. In 2010, Anoka will be a community which:

- Creates an environment that invites and attracts business and industry to the area through effective transportation, utilities, communication systems, and other accommodations;
- Realizes the economic benefits of river front development and redevelopment;
- Is open minded and aggressive in exploring options for development that will not only benefit business but will also address the long—range needs of the community;
- Encourages and promotes business enterprises which provide needed products and services not only to the people of the area but also to populations and business throughout the world;
- Provides a variety of rewarding employment opportunities for the people of the area;
- Promotes and supports the efforts of businesses to maintain viability, to change, expand, or improve services and products; to improve efficiency and competitiveness; and to operate in an environmentally sensitive manner.

VALUE OF THE NATURAL ENVIRONMENT

Anokans value the natural environment, recognize the unique resources represented by their rivers, and enjoy the many opportunities which such resources offer. In 2010, Anoka will be a community which:

- Appreciates the economic and recreational value of rivers, lakes, and wetlands while respecting and protecting native wildlife and valuable habitat areas;

- Strives to operate with a minimum of pollution and waste, encouraging residents and businesses to participate in waste management programs that are both effective and convenient.

SERVICES FOR QUALITY OF LIFE

Anokans recognize the importance of many services in maintaining their quality of life. These services will continue to be important to them, and in 2010, Anoka will be a community that:

- Has transportation systems that are modern, accessible, well maintained, low polluting, and energy efficient;
- Has a road and bridge system which supports development opportunities;
- Recognizes the importance of parks and recreational facilities which are clean, safe, well-managed and accessible, and develops properties with consideration for such facilities;
- Is committed to a high-quality educational system which makes the most of its physical, financial, and human resources to provide educational opportunities for all citizens;
- Promotes the physical and mental health of all its residents, and supports the industries, professionals, and agencies which provide such services.

—Reprinted from "Building Bridges to a Better Community," Bill Mittlefehldt, editor. Contributors: City of Anoka, Anoka Area Chamber of Commerce, Anoka High School, Anoka-Hennepin District 11's Community Education Department. Bill Mittlefehldt can be reached at Anoka High School, 39397th Avenue North, Anoka, MN 55303.

The two student visions below are rougher versions sent to us by teachers who used *Exploring Sustainable Communities* in their classes. Obviously there are items listed that any teacher might want to query and discuss further. They definitely bring out the concerns and the aspirations of the students.

JOHN JAY HIGH SCHOOL, KATONAH-LEWISBORO SCHOOL DISTRICT, NEW YORK

The Katonah-Lewisboro School District lies at the outer reaches of the commuting area to New York City, in Westchester County, an hour's ride by train. Nearly everyone depends on the city in one way or another. John Jay High School and its surrounding community are undergoing dramatic changes. Students walk through construction areas daily as they move from one class to another.

A Community Vision for Katonah-Lewisboro by Senior High School Students Teacher: Dick Parsons

People

- A mixture of ethnic and age groups
- Helping and friendly population
- People are environmentally aware.
- Population growth is a direct result of space available.
- A separate senior citizen community

Housing

- No condominiums; strict zoning
- No large developments
- Limited number of housing complexes and condo's
- Well-spaced housing
- Separate houses within walking distance of small towns

Schools/Libraries

- Separate elementary schools; combine middle and high schools to give feeling of community
- Buses for all students; parking available for all students
- Set number of kids in each grade
- There are public agricultural schools with good libraries
- Local town libraries

Jobs/Business

- No chain businesses
- No large industries—rather family businesses
- Businesses are in good competition: no monopolies
- Unions for all workers who want them
- Commercial/residential areas separate; no houses in commercial areas

Health Care

- Local medical centers; centralized hospitals
- Local medical centers for residents only
- Family doctors for personalization

- Family physicians for homes; hospital no further than next town
- Local child care units

Crime

- Neighborhood watch for citizens
- 16 and over may be jailed
- There is a feeling of trust in community.
- Follows the golden rule
- Nice/personable police officers

Transportation

- Local trolley down main road
- Trains connect to large city and local towns
- Public transportation for schools
- Shuttle bus system
- One car per family rule
- Bus routes have loops that are regularly followed.

Amenities

- Town festivals and celebrations
- "Get to Know Neighbor Day"
- Sidewalks
- Big "natural park" separate from but connected to town park
- Many smaller playgrounds

Environment

- Community clean-up once/month
- Deer season year around
- Natural reservations set aside to maintain natural environment
- Wildlife sanctuaries
- Clean, preserved wooded areas

Public Involvement

- Everyone is involved in community; no one excluded
- PTA; neighborhood watch
- Town meetings for significant changes in town; everyone welcome
- Vote on anything affecting the environment

Other

- Driving is at parental discretion.
- Maintain population at sensible level
- Use of hydroelectric/solar energy
- Open land to be used as a "common space"

A Community Vision for Rockville, Maryland. Teacher Marty Creel, Interdisciplinary Specialist for Social Studies, Montgomery County Public Schools

Rockville is a northwest suburb of Washington, D.C. Many residents commute to Washington, but more businesses and industries are coming to Rockville and the surrounding region. The "Interstate 270 Technology Corridor" passes through Rockville. The city is growing fast, and many of its developments are fairly new. Residents are very vocal and active in their city's development. The city has recently sponsored a visioning process, "Imagine Rockville."

People

- Anybody of any race, religion, ethnicity, or income level can live in our community.
- There are more young people, but it is a community with youth, adults, and elders.
- Our community will not have many families that move away.
- There are no new housing developments, because the population is slightly decreasing.

Housing

- Our houses will be regular houses and townhouses in clusters of about five. The families in each cluster will work together and help each other out with maintenance, day care, etc.
- The newest housing developments are replacing older ones. Abandoned shops and houses are being demolished and apartments are being built. Those are the only developments being built.
- Every year all the neighbors will get together and repair all the old houses (new wood, paint job, etc.)

Schools/Libraries

- Most children attend public schools, which are as cool as private schools, but are free. Teachers are well-educated, and students are disciplined.

- My school is within walking distance. I can walk or bike to school.
- Middle school starts at 9:00, because adolescents need their sleep.

Jobs/Business

- People have many different jobs. Some work for the government, and some work for themselves.
- Most businesses are located in concentrated non-residential areas.
- The unemployment rate is dropping due to new opportunities for people who didn't have them before.

Health Care

- Health care—for everyone—is available through a comprehensive government program.
- People will apply for it from a Web page on the Internet.
- Health care will be close. Period.

Crime

- People will trust each other and feel very safe.
- To prevent crime, police will monitor, and it will be emphasized greatly in school.
- A kind, efficient police force limits the urge to commit crimes.
- All criminals will be tried, given one chance to repent, and if guilty of another crime, shot dead immediately.

Transportation

- Most people get places by foot, bicycle, or public transportation. There will be sidewalks on every street, allowing people to walk and ride bikes without feeling like they will be hit by a car. Only one car is allowed per family, and each person is allowed to drive only so many miles per year.
- There will be a complicated system of metro subways, planes, helicopters, buses, gravi-cars, and hover craft.
- Solar energy will be used for energy for transportation.

Amenities

- For daytime recreation, we go to the community center, or the abundant and beautiful parks. Playing fields are also available.
- Historical tradition, culture, and rituals will be preserved in ethnic recreation centers. There will be a celebration for every holiday celebrated by 5% of the citizens.

- There will be carnivals, fund-raisers, fairs, get-togethers for holidays such as 4th of July, and fairs promoting subjects like art, literature, the sciences, nature, etc.
- There will be a giant mall with a roller coaster in it like the one in Minneapolis, and it will be located in Silver Spring, Maryland.

Environment

- Forests and fields will be preserved. For every section of natural land that is depleted, another is saved somewhere else.
- People in the community are required to do some environmental work, be it cleaning the local river or picking up trash in the park. Since cars are run on electricity and natural gas, the air is not polluted, nor the water.
- Anyone caught dumping, littering, or polluting our water will be fined a minimum of $2,000.

Public Involvement

- Every citizen will pitch in their part for the good of the community.
- There will be weekly meetings for citizens where they can voice their opinions on changes and take part in local government.

Other

- Citizens will have private areas where they may separate themselves from the community through meditation.

As these examples show, working with youth to develop a vision for their community's future is a valuable educational opportunity, linking knowledge in many subjects, and encouraging student creativity.

CHAPTER 10

Facilitating Education for Sustainable Development

Jean Perras

THE STRATEGY OF A CANADIAN NON-GOVERNMENTAL ORGANIZATION

In 1991, as part of a response to the Report of the United Nations World Commission on Environment and Development (Brundtland Report), a cross section of Canadians, representing education, government, business, environmental and Native organizations, established a unique non-governmental organization called Learning for a Sustainable Future. The mandate of the new organization was to integrate the concept and principles of sustainable development into the formal education systems in Canada. This is a brief overview of how it has carried out its mandate to date.

In 1991, few people had any but the vaguest idea of what sustainable development was or had even heard the term. The first challenge for the Board of Directors of Learning for a Sustainable Future (LSF) was to reach an understanding among themselves on what sustainable development might mean and what it might imply for education. The Board began to realize that no clear definition was possible, that sustainable development was a complex process of questioning, learning, planning and change, not a destination but a journey. Moreover, the process had to involve all sectors

Jean Perras, Executive Director, Learning for a Sustainable Future, Ottawa, Ontario, Canada.

of society, not just education. The implication of this perception was that LSF abandoned ideas to develop a program with a set of aims and objectives and a step-by-step method of achieving them. Instead, LSF defined its role as setting the contexts for inquiry and engaging Canadian educators and communities in a process of planning and change which was at the core of a move towards sustainability. LSF would try to facilitate this process of lasting change within the education systems of Canada, one that would affect all levels of the education system from policy makers and curriculum writers of Ministries of Education, to teachers and students in the classroom.

This strategy of change had four main elements:

consultation to seek the advice and support of Ministers of Education and their key personnel, teachers' organizations, business leaders, and government officials;
facilitation of planning for strategies of education by key stakeholders in provinces and territories;
encouragement and support of provincial programs of teacher education;
development of appropriate education materials to allow teachers and students to engage in the process of inquiry, research, decision-making, consensus-building, and appropriate personal and cooperative action for change.

To outline a program approach is one thing but to implement it is another. The Board faced two major challenges. The first is that there is no central education system in Canada. The British North America Act of 1867 gave responsibility for education to Canada's provinces so that in 1991, strategies had to be developed for ten provincial and two territorial departments of education, each with distinct curriculum approaches, teachers' federations, and school board structures. The second challenge was for an outside organization, such as a non-governmental organization, to gain the credibility needed to suggest new methodologies for any aspect of education, especially curriculum design/development.

There were also, however, some unique opportunities. The first was a long tradition of environmental education in Canada supported by an active NGO community and strong public awareness. The second was the twenty-year-old federal government-funded program of development and global education, by which many non-governmental organizations had been able to create an awareness among thousands of teachers of global international and national issues of development, environment and society, and innovation in teaching about them. Education for sustainable development was, in a sense, an extension of environmental, development and global education, bringing the added dimension of economic development and resource conservation into the perspective.

The third major opportunity was political support. Canada had embraced the idea of sustainable development and set up the National Round Table on the Environment and the Economy, the equivalent of the US President's Council on Sustainable Development. In addition, most provinces and territories had created their own round tables on economy and environment to look at issues of sustainable development on a regional basis. Many of these had published a provincial plan for sustainable development, and most plans included a strong statement on the crucial importance of education in achieving it. For example, in one province, New Brunswick, the report by the Round Table, signed by the Premier, stipulated that education for sustainable development was to be a priority of education and integrated into all areas of the curriculum. In addition, polls had indicated very high public concern for issues of the environment, fueled by publicity about the United Nations Conference on Environment and Development in 1992.

Against this background, Learning for a Sustainable Future began its first rounds of intensive consultation, introducing the organization and its objectives to Ministers of Education and their officials, teachers' federations, business leaders and representatives of government across Canada.

To establish its credibility in the field of formal education, it was essential for LSF to define its rationale and vision of education for sustainable development, one that was geared to existing curriculum requirements, one that would be recognized as an innovative approach to educational change based on sound and excellent pedagogy, and one that could be recognized, accepted, and adapted by teachers and other educators across the country. To do this, Learning for a Sustainable Future developed a comprehensive list of skills, knowledge and values that would be needed by Canadians as they began to face the changes needed to contribute to a sustainable future.

This was a unique exercise. While curriculum designers routinely set out the knowledge, skills and values expectations for specific subject areas, this was perhaps the first time that these suggested outcomes transcended subject disciplines to describe the requirements for the whole future citizen. In a sense, it set out the objectives for all of Canadian education, a philosophy geared to the challenges of the 21st Century.

Learning for a Sustainable Future circulated its knowledge, skills, and values list across Canada to about 800 leading educators, government officials, business leaders, non-governmental organizations and others, and correlated large and thoughtful responses into a final document that might be said to represent a consensus among Canadian educators on what is needed today to educate a citizen of the world of the 21st Century.

With this validation, Learning for a Sustainable Future entered into its wider consultation with provincial and territorial stakeholders to identify regional strategies of education.

Each of these stakeholders' meetings involved in-depth discussions to gain the acceptance and participation of teachers' federations, Ministries of Education, school boards, and political and business leaders. Teachers' federations with established global education committees hosted most of the meetings, sent out the invitations and helped Learning for a Sustainable Future to facilitate the agenda. Moreover, they gave the project credibility among provincial and territorial educators.

These first stakeholders meetings were crucial in that they brought together, often for the first time, representatives of all sectors of society. Labor, business, and corporate executives from different government agencies, non-governmental and Native organizations, teachers, administrators and parents examined the changes that were happening to their environment, their economy, and their society in general, and came to terms with the need for change. They formed advisory groups to plan cooperative strategies for education that had the acceptance of the public as well as the support of the formal education system, the essential elements for lasting change in education.

Provinces and territories have progressed at different rates according to the changes in government and educational policies, school board structures and curriculum priorities. However, today there is a pattern of established programs of teacher education and materials development, spear-headed by teachers and strongly supported by other stakeholders. Recently, the Council of Ministers of Education, Canada (CMEC), produced a report entitled: *Educating for Sustainability: the Status of Sustainable Development Education in Canada*. The report can be found on CMEC's web site at: www.cmec.ca.

In the meantime, Learning for a Sustainable Future developed a number of key materials to help clarify the processes of curriculum design, and classroom activities and methodologies. These included the *Cross Curricular Teachers' Guide* and an introductory workshop, *Toward a Sustainable Future*, designed to help teachers understand sustainability and its implications for education, and a series of case studies called *Inquiries into Issues of Sustainability in Canada* and *Awareness-Raising Tools for a Sustainable Future* such as the collapse of the cod fisheries, sustainability of fresh water, etc. All materials were tested in a cross-Canadian program of teacher workshops that continues today, and all materials are offered free of charge on the Learning for a Sustainable Future web site (*http://www.schoolnet.ca/learning*). Almost 3,500 teachers have been reached by the workshops.

One of the most important measures of Learning for a Sustainable Future's progress has been requests by the Ministries of Education for advice on the design of new curricula in Canada to help to ensure that the perspective of sustainability is included in new policy documents. Learning for a Sustainable Future has contributed to the development of the common curriculum in science and social studies in the Atlantic provinces; to Ontario, Alberta and Manitoba science curricula; and to the Pan-Canadian Framework for Science Learning Outcomes. Today, much remains to be done, especially in the field of teacher education, but the foundations for learning for sustainability have been laid in new curriculum documents across the country. What was an unknown concept in 1991 is now becoming a familiar context for classroom learning, as well as policy-making and curriculum development.

The work of Learning for a Sustainable Future has received international attention and recognition. In 1996, it formed alliances with the Commission on Education and Communication of the World Conservation Union (CEC-IUCN) to follow up on Chapter 36 (Education) of Agenda 21. Both organizations helped the Québec Teachers Federation organize the Planet'ERE Forum on environment and sustainable development education held in Montréal in November 1997, attended by more that 700 educators from 35 Francophone countries. They also developed a model for legislation on sustainable development education, which was presented to the parliamentarians of the Americas in Québec City in September 1997, and at the meeting of the Arctic Council on Sustainable Development in Whitehorse, Yukon in 1998.

The Learning for a Sustainable Future strategy is now reaching school systems in other countries through its bilingual (French and English) web site. Learning for a Sustainable Future is proud that its work has been recognized by EXPO 2000, the first World Exposition of the 21st Century, which will take place in Hannover, Germany from June 1st to October 31st, 2000. The theme of EXPO 2000 is *Humankind—Nature—Technology*. Learning for a Sustainable Future will showcase its sustainable development education model and its Web site under the EXPO 2000 theme, *Projects-all-over-the-World* in the Canadian Pavillion.

Education for a sustainable future is a growing world movement. As the 21st Century presents its fundamental challenges of accelerating change in society and economies, increasing interdependence, and the urgent search for more sustainable ways of interacting with the environment, education must rise to the challenge and assume responsibility for the preparation of present and future generations. As experience in Canada has shown, the content and concepts of education for a sustainable future encourage the knowledge, skills and values needed by the workforce of the changing global economy, by communities struggling for self-sufficiency in an age of declining resources, and by individuals looking to maintain a personal identity in an increasingly technological age.

The challenge of effecting change in society or in education begins with a vision of hope and the recognition of the changes that must take place to make the vision a reality. Above all, the lesson to be learned is that progress towards sustainability is a cooperative venture.

CHAPTER 11

Transforming Public Education
Sustaining the Roots of American Ideals, Our Economy, and Our Environment

Richard Benjamin and Susan Hanes

> "The buzzword is 'sustainability,' and we're already getting tired of it, but the idea makes sense . . . We no longer simply need development: we need better development . . . Growth should be driven by quality and tempered by the long-range interests of the people who live here . . . Sustainability is precisely for people, including the people who come after us." (*Fort Myers News*, July 28, 1998, editorial)

The quality of our life and the future of our American ideals, depend to a large extent on the quality of our public schools. Soon, students will need to master the traditional basics earlier, learn to apply them in rapidly changing real life settings, and go on to better acquire the additional basics that will be required in the future—learning how to learn and loving learning, problem solving in applied settings, collaboration and teamwork, and character traits such as respect, resiliency, and responsibility. Currently, many students are not served well. We know this from tests scores as well as from

---

Richard Benjamin, Superintendent, Cobb County Public Schools, Marietta, GA. Susan Hanes, Director of the Metro Atlanta P-16 Council, Georgia State University, Atlanta, GA.

other, more authentic performance assessments, from "benchmark" information, from tracking college freshmen, and from on the job performance reports. We know students at all levels can learn more and we are becoming acutely aware of new demands that current and future students must be prepared to meet. And, it seems that these new demands are emerging at an increasingly rapid rate.

In our country, for the sustainability of our national ideals, our economy, and our environment, all citizen stakeholders, including the educators, must become active reformers of public education.

School reform is difficult. The issues are complicated, controversial, and respond slowly even to concentrated efforts. The vastness and importance of the task are obstacles. Another obstacle is the very success of current practices for many children on many educational objectives. The energy for reform is diminished by the success that is achieved in some ways, for some students. There is, of course, evidence that our public schools do well for many students, and that the schools are improving. However, they not only need to improve, but they must improve at the process of improving and do so at a rate equal at least to the rate of change evidenced in our society. As the noted researcher and author, Diane Ravitch (1985) asks, "Why shouldn't all children get the best education we can afford to provide?"(p. 4). In fact, she believes the future of school reform depends on whether the school reformers maintain such a commendable democratic ideology and whether the reformers forge an indissoluble link between quality and equality.

To forge that link, public school reform must borrow heavily from the literature and best practices of sustainable development and of learning communities, including systems thinking. This means more emphasis on long term, clear purposes, high expectations, and on transformational strategies, which build on current best practices, with a commitment to continuous improvement through continuous learning. If a student receives a basic, balanced education, the habits of learning and thinking will become a lifelong pursuit.

To sustain means to maintain or cause to continue in existence or in a certain state, or in force or intent. Sustaining our national ideals, our economy, and our environment are worthy goals in and of themselves, but each is in turn an interrelated part of a complex system. As parts of a complex system, they are related in intricate ways and are individually and collectively supported by the quality of the educational system. Our educational system must work and work very well for us to sustain the quality of life we have and to explore improvements in it for us as well as for future generations. Quality public schooling will be decisive, and it will need to be a careful blend of the best of what it currently is and a commitment to continuously improve. Public schools will either reform, or transform, or will not be sustained and not be available for their important work in sustaining our democratic ideals, our economy, and our environment.

It is no accident that the logo for the National School Safety Center (1999) is an American Flag in the shape of an apple. The center literature points out that: "School Safety—(is) a national concern (because) America's system of public education plays a key role in developing our nation's youth into knowledgeable, responsible, and productive citizens. It contributes substantially in preparing young people to participate in this country's democratic system, a government by the consent of the people" (p. 1).

The center makes the relationship clear, and a threat to public schools, to free, high quality education for all students, is, in our country, a threat to our continued adventure in democracy. There is a special importance of a high quality educational system in this country because of our national purpose.

Mortimer Alder (1988) makes this clear by pointing out that since the people have replaced the prince and are the self-governing rulers of the republic, it is important that all students receive a high quality preparation for the duties of citizenship. Preparation for citizenship, for earning a living, and for discharging one's moral obligation to lead a good life and make as much of one's self as possible, constitute, for him, the three key objectives of any sound system of public schooling. And, he feels the present public school system does not serve any of these objectives well enough.

Progress toward preparing students for all three of those roles will require progress on three types of goals—the acquisition of core knowledge, the mastery of core skills, and the development of core values or character traits.

In a related way, Heidi and Alvin Toffler (1995) feel that the shift to an Information Age civilization, the accelerating rate of change, and the implications of both together are serious concerns to those who favor the sustainability of democratic ideals. The same factors pose specific challenges to the sustainability of our economy and our environment. In turn, risks to these eventually pose new threats to the sustainability or continued development of democratic ideals and practices.

SUSTAINABILITY OF WHAT? SYSTEMS HAVE PURPOSES

This is a country like no other, born of clear democratic ideas and ideals forged into national purposes. American values—individual liberty, freedom, justice, and equality under the law—are some of the highest and most cherished ideals ever expressed. They have guided our own adventure with democracy and have served as beacons to freedom-loving people everywhere.

In order for our adventure in democracy to continue to grow and improve, we will have to provide a free, high quality public education for every citizen. This education must address not only the core-content knowledge and core-process skills emerging as important, but also must address the core-values and character traits needed historically, needed now, and needed more than ever in the future.

Sustainable development—meeting the needs of the present without compromising the ability of future generations to meet their needs—leads us to consider a common commitment to the public good. It also leads us to consider what exactly is worth sustaining. The concept of sustainability can be, and is, criticized for its vulnerability to a lack of an ethical underpinning. Certainly there are current practices that are unworthy of being sustained. Gender discrimination and multicultural issues come to mind.

However, a commitment to the common good, to civility, and to responsibility, are ascending as prerequisites to our ability to sustain democracy. Similarly, the future protection of our natural environment and of our economic stability in the information age also put demands on the public schools of America, which, if ignored, will weaken both our public schools, and, ultimately, the democratic ideals the public schools serve. This paper presumes the human goals of democracy, and the economy and environment to support them, and an educational system to support all of them, are worth sustaining. These clear purposes can and should guide our "system" improvement efforts.

America's students will grow up to live in a world characterized by increasingly rapid change. To serve individual students, and the nation, the public schools of America must transform themselves, or be transformed, into learning communities focused on continuous improvement. The students, and the nation, will be better served if the public schools prepare students who, as individuals, meet their own worthy needs and those of their families, earn a living, and, as citizens, foster the sustainability of our national purposes, our natural environment, and our economic growth.

Business leaders are recognizing that knowledge becomes the ultimate substitute as the central resource of an advanced economy because it reduces the need for materials, labor, time, space, capital, and other inputs. And as this happens, the value of information and knowledge-related skills soar (Toffler, 1995).

As the value of information increases, the value of employability skills related to the production and distribution of knowledge increases. Therefore, the public schools of this country must be a more effective part of shaping the workforce of the next century toward national leadership in an information-based world economy.

Beyond threats to our national ideals and our future economy, the current and future threats to our environment are growing to the point that decisions made in the next few decades may be decisive for the quality of life on earth. According to McKibben (1998), hunter-gatherers used 2,500 calories of energy a day (all of it food), whereas modern Americans use seventy-five times that. He reports that a modern human being uses 31,000 calories a day, most of it in the form of fossil fuel. The average American uses six times that amount of calories. Human beings, unlike deer, can eat almost anything and live at any level they choose, and can import what they need from thousands of miles away. And human beings, unlike deer, can figure out new ways to do old things (p. 57).

It is precisely this task, preparing students to figure out new ways to do old things that will require the public schools of this country to transform themselves. This is necessary to prepare students to achieve sustainability—sustainability in our national purposes of liberty and justice, sustainability in our emerging information economy, and sustainability in regard to our natural environment.

The discussion of school reform in this country seems to have waxed and waned over the decades, but generally continued unabated since the earliest times when the very need for public schools was vigorously argued. The same, or similar arguments, are made today as reform is discussed, but the emerging demands on public schools, with the emerging alternatives, makes the situation more urgent.

A DISCUSSION OF AN EMERGING TRANSFORMATION OF A PUBLIC SCHOOL DISTRICT

What follows is a discussion of elements of the emerging transformation of public schools-building on a solid foundation of quality while attempting to become what they must be for current and future students. This discussion is offered in an effort to invite widespread involvement in educational improvement by making explicit certain assumptions and views of the improvement process.

This discussion will progress on the assumption that free, high quality, public schools are necessary for America to continue on its journey to live out and extend democratic ideals. A second assumption will also guide the discussion, that current public schools in America must undergo transformation in a short period of time to be as effective and efficient at preparing students for the future as the times will require.

A third working assumption is that the transformation can and must come from a balance of internal initiatives in partnership with external, entrepreneurial, individuals, and agencies. The assumption being that "sustainable" transformation will come with genuine internal reflection and a desire for continuous improvement, and, from an authentic desire and need to improve customer satisfaction through strong community partnerships. The "customer" is the community. The community pays taxes in order to receive a "result"—a well-educated individual, future citizen, and member of an international competitive workforce. This will be the source of necessary tension and urgency, learning from and setting an example of what other sustainable organizations will need to do.

Illustrations used in the following discussion come from the practical experience of schools, school systems, educators, families, and community leaders actually engaged in systemic transformation of public schools. Many examples come from recent school improvement efforts in Cobb County, Georgia. The examples reflect the results of sometimes-extensive participation of staff and community in efforts to create, identify, and assess personal values and mission, and to blend these toward a shared set of values and a shared mission. This means the details, especially from school to school can and will vary, and that practicing democracy at the same time we seek to strengthen it, will be troublesome. Honoring a diversity of opinion will create a diversity of solutions. What remains as stable are the goals, with flexibility of methods. This emerges as a key point in the transformation of schools and is a key tenet of those like Harvey Silver who urge, "simple and deep school improvement methods." (1999)

The future of America, our children and our ideals require our best efforts. The above discussion of "why" public school transformation is important needs to be kept firmly in mind because addressing the "what" and "how" of school improvement is complex and difficult to the point of painful, and nothing less than large-scale, sustainable transformation of public education will do.

REFORM OR TRANSFORMATION: THEM OR US

Educational reform isn't going well—either in terms of quick, simple reform, or in terms of sustainable transformation—continuous improvement. This may be because no silver bullet or simple answer exists. And, top down "reform" of anything seldom works other than in the short term.

More likely, it is not going well because we humans all too often start reforms with a list of changes for someone else, and seldom start with the only person we can really change—ourselves. Successful transformation is needed and will need to be more "us," all of us, and less of how to change "them." Transformation really is a "system" issue, and slight improvements in one part, our part, will more likely lead to a meaningful change in the other parts and in the key relationships among the parts. Each and every one of us *must* accept the necessity of change and our own responsibility FOR IT.

Also, unless the focus is on students and actual learning for all students, our efforts will continue to fall far short. The big paradigm shift for public education in our time is the shift from "coverage" and "seat-time" to learning and results. Being present will no longer be a substitute for learning. As a drive for "learning for all" replaces ideological and methodological reforms, public education and educators may also regain deserved understanding and support.

To a surprising extent, a sense of urgency for effective reform is missing, in spite of strong rhetoric, and a near panic for improved results. It seems that all too often public officials and others would rather have the issue of educational reform to stir emotion than to have the results of meaningful, sustained improvement. Effective reform will require levels of collaboration, sacrifice, and continuity not recently achieved on this topic by elected officials so far.

This lack of interest in much other than blame and short-term solutions is a problem. This is compounded by the complexity that current school reform efforts need to attend not only to the inadequacies of the past and the emerging needs of the information age, but also to the next "age" which will come all too soon. Theories about the next "age" abound, perhaps led by the knowledge that the cycle of "ages" seems to be shortening dramatically. The fact that the rate of change is itself accelerating was a key point in an earlier Toffler book, *Future Shock* (1971). Therefore, a key question will be about the nature and duration of the information age. Will it follow the shortening cycle of the agricultural and industrial ages or will it represent a new and more enduring sort of transformation? These ages are often characterized by the cycle of jobs—over 90% in agriculture at the start of the industrial revolution diminishing to under 10%. Will a new age reveal itself when 90% of the population is working in knowledge or information jobs, and if so, what will characterize that age? The sustainability of our democratic ideals, our economy, and our environment will benefit from clear monitoring and analysis of these issues. Educational transformation will require it.

Perhaps "reform" as we think about it isn't as good an idea as transformation—acknowledging need for dramatic structural improvement, carefully building on current strengths. Perhaps the butterfly is a better metaphor—a healthy caterpillar undergoing a careful metamorphosis into something so completely different as to belie the earlier incarnation.

Students, their roles, and their family's roles, can, and probably should, be the center of what is thought of as "school reform" in this country. Students taking responsibility for their own learning and the maintenance of a learning community, and families taking a more direct role in facilitating the learning of their own children, ought to be a focus of the highest priority in the improvements to be made in our "system" of public education.

Systemic, comprehensive, sustainable transformation focusing on the individual student and carefully attending to the relationships the student has with his or her family and teachers and the relationship between the schools and the community will make the big difference. This will guide American public education in the direction of sustainable school transformation because it will strengthen first the individual, then the family, the school, the community, and the nation.

Our students know this, and listening to them, then reflecting on what they think and what they feel about the future, is mostly encouraging and often inspirational. Most students are well aware of the direction and the pace of change that will be dominant throughout their lives. They certainly aren't "getting tired" of sustainability or the idea that current generations need to be more responsible for long-range interests. Students do often seem impatient with "adults" who seem attracted by short-term, simplistic, and obviously selfish solutions to the complex problems we hold in common. Such approaches threaten the sustainability of the quality of their life, and our students know it.

Most students actually want more challenging schoolwork and see the need for better ways of teaching and learning. They often know in real detail what is coming and what needs to be done to improve the schools they attend, for themselves and for the students who will follow them. Most seem ready to accept good examples set by adults who start reform with themselves and see the need to reject "adult" models from people who really haven't grown up. Hence, the basis of real transformation is the individual student.

THE WHAT AND HOW OF PUBLIC SCHOOL IMPROVEMENT: WHAT ARE THE PARTS AND HOW ARE THEY RELATED?

Leading and managing, for transformation, for sustained school improvement, will require the establishing of learning communities and a comprehensive, or systems approach. Systems' thinking is a clear prerequisite. This requires knowledge and understanding of all of the parts of the public school system, how the parts are related, and, most importantly, the purpose of the system. Clear worthy purposes, profound knowledge of the systems, skill in managing transitions, and resiliency and tact as character traits all come into play.

What is the purpose of the system, and what are the key parts of the system? According to Deming (1995), a system is a network of interdependent components that work together to accomplish the aim of the system. Deming is clear that a system has an aim. The aim, the purpose, the "why" of public schools and the why of the need for transformation has been addressed above, and for our purposes focus on democratic ideals, our economy, and the environment. Comprehensive transformation means attending to all necessary components at the time needed. It does not mean thinking a large chasm can be jumped with a series of small steps. Sometimes a lot of things, seemingly different steps, have to happen nearly at once for improvement to be successful.

Unfortunately, even high levels of time or energy may not help student achievement if it is not applied to points of greatest leverage—to the high yield strategies. Systems thinking helps identify if, and how, a component is actually contributing to the system goal by requiring clarity about how parts or components are related to each other and to the goal.

Being more accountable means being more responsive, more likely to monitor change and progress, and more likely to make appropriate adjustments or improvements. In this sense accountability leads one toward improvements and sustainability. Public school systems have all too often been characterized as patriarchal and paternalistic, putting control, consistency, and predictability above results. As unresponsive bureaucracies, immune to feedback, they have lost much of the public support they need to function. This has caused concerned people within and outside of public education to explore alternatives to the current relationships. New forms of "public engagement" are being considered. Lessons from other types of organizations are being learned and applied.

The business sector, with the successes achieved by greater attention to customer satisfaction, and, the public schools, with knowledge and expertise about learning and knowledge, have perhaps more to contribute to each other than at any time in the past.

For public education, parents and students, and the wider community, are seen as necessary actors in the system as well as customers.

Students especially will play a key role in transformed public schools. Unless we find ways to work with students so that they take more responsibility for their own education, and their own behavior, we will continue to work hard and not see the results that can and must come about.

Self-assessment especially will help students (and teachers) to become more active partners in the learning / teaching relationship.

Relationships based on partnerships, empowerment, and stewardship are starting to replace those based on hierarchies with the "head" doing all the thinking (Block, 1996). Both Block and Covey (1994) advocate "stewardship" contracts or agreements, to insure that "each party has to struggle with defining purpose, and then engage in dialogue with others about what we are trying to create" (Block, p. 29). "The stewardship agreement creates a synergistic partnership to accomplish first things first together"(Covey, p. 224).

These considerations bring together key concepts of community and stewardship in highly practical ways. Negotiating improved relationships among all the key stakeholders is fundamental to the work of transforming our public schools.

THE BIG PICTURE

What is the big picture? What are the large transformations or shifts that characterize the improvements underway and those that will be needed?

First and most importantly, is a paradigm shift from process or "coverage" to "results," with results defined more broadly, beyond "test scores." No matter how this is expressed, whether as a move away from "seat-time," or process, or even away from an emphasis on the act of teaching, it means that learning is the focus, and if learning does not take place, all the rest must be reconsidered. Strategically, it means a system focus on improving student achievement, broadly defined, to explicitly include realistic application of knowledge, skills, and character across subject areas and in more real life settings. While standardized tests will provide initial measures, the emphasis must shift rapidly to more real world products or exhibitions.

Second, the big picture means a paradigm shift from identification and implementation of programs and knee-jerking fixing of things when they are obviously broken, to continuous improvement. Strategically it means effective and efficient use of all resources, using the tools of continuous improvement and inventing new tools. It means careful application of the lessons learned about human beings and the change process. It means building systematically on the greatest strength of America's public schools, the knowledge and experience of teachers in creating and maintaining a learning organization—where everyone understands they are engaged in an activity that has meaning for them, and that they, as individuals, matter.

Third, the big picture means a paradigm shift toward more responsive, more interactive partnerships with increasingly informed stakeholders.

It means having all the stakeholders in on defining direction and assessing value. Strategically, it means better listening, communication, and responsiveness to all stakeholders, more accountability—defined and implemented in productive ways. The stakeholders, of course, are all of us in the community with a vested interest in an educated youth participating in the sustaining of our democracy. The shift from simplistic, narrow accountability models amounting to little more than "blame" searches to meaningful comprehensive, productive mutual accountability models is key. Schools are not really being responsive or accountable when they implement mindless accountability models, which are so shortsighted and simplistic as to actually do harm. An important transformation requires shifting from judging themselves, generally at the end of the year with narrow test scores, to using a wide variety of measures, emphasizing actual real world performance and external customer satisfaction, with the explicit recognition that public schools have several key stakeholders.

These paradigm shifts are almost always represented in public school strategic plans and, together, form the basis for the following analysis of sustainable school improvements.

PARADIGM SHIFT #1—FOCUS ON RESULTS—THE HEART OF THE MATTER—INCREASE STUDENT ACHIEVEMENT

The shift is from "covering" the curriculum to "learning for all" and the emphasis is on results and the frequent monitoring of results and making the frequent adjustments needed to insure success for all students. The measure of progress in improving student achievement will likely start with standardized test data and evolve to more meaningful, more timely measures which focus more on the application of knowledge, skills, and character in real life settings.

We must keep the focus on learning and results and remember that we are all students. Focus on learning means the development of accurate, timely, authentic measures of progress and generating useful feedback. It has been said, "feedback is the breakfast of champions," and indeed feedback is a key element in systems thinking, learning, and systems improvement. When one is determined to understand not only the parts of the system, but the relationships among the parts, understanding and improving feedback is key to understanding and improving the performance of the whole system. By "feedback" we mean, of course, gathering and using information about results in terms of student progress, and opinions, comments and suggestions from students, other internal "clients" and the community. Frequent monitoring and continued adjusting is the opposite of covering material and then simply moving to the next topic or simply installing another program.

The original effective schools research started with a clear picture of the then desired results. An operational definition was crafted—equal proportions of sub-populations demonstrating mastery of basic skills as measured on standardized tests. Schools considered to be effective had roughly equivalent achievement levels for black/white, rich/poor, or male/female populations demonstrating such mastery. The data had to be disaggregated so the low achievement of one population was not lost in the average achievement of all taken together. The method was a powerful tool for improvement. It focused on results and clearly proved useful in helping people in identifying the things that "effective" schools had in common—the correlates of effective schools. The correlates are the practices and policies that appeared to produce student success.

Madelyne Hunter (1982) used a similar but less formal approach in identifying effective classroom practices. Now, we see research-based rubrics, which embody classroom "best practices," supported by data. Some grow out of semi-informal "action research," and some are based on formal research using experimental designs. *Enhancing Professional Practice: A Framework Enhancing Professional Practice* (Danielson, 1996) is a good example of the latter.

The same approach was used to identify best practices at organizational level by Peters & Waterman (1982) when they identified the lessons from America's best-run companies.

The approach, while powerful and clearly useful at the classroom, school and organization levels, has limitations. In the case of the original effective schools research, it focused only on one type of educational goal (performance on standardized tests), it established only a correlation, not a casual relationship, and it depended on a keen eye to discern exactly what might actually be making the difference—even when the classrooms, or schools, or companies are accurately separated into "high" and "low" performance groups.

Lawrence Lezotte made a good point in recent discussions that one must learn to discern what the leading indicators might be and not rely only on the trailing indicator of achievement results. He points out that identifying the leading indicator "best practices" or essential principles of human learning serves school improvement thinking better than only looking at the trailing indicator—student achievement. The coveted Baldridge Award also turns our attention to "approaches" as well as to results when one is interested in learning to identify best practices.

It takes a conscious effort to get past effectiveness as defined as equity on standardized tests to excellence. Excellence means more than performance on standardized tests and more than simply "equal proportions" performing at what might be low levels of equal achievement. The future will demand high levels of functioning, of all students, of a variety on assessments beyond standardized tests. Equal proportions of learners at a low level of achievement are not anyone's ultimate goal. Further, there are important educational outcomes not measured on standardized tests.

Nonetheless, the approach has been an essential step. When paired with proven improvement strategies such as described in *Sustainable School Reform* (Lezotte & Jacoby, 1992), a school or a school system has the beginnings of an empowering approach to continuous improvement. When reinforced, as it is in Georgia by an "Effectiveness Review" process, which is grounded in the effective schools correlates, a school staff and parents have some powerful support for carrying out their responsibilities. Identifying and supporting low achieving schools through this process of observations, interviews, and analysis of achievement data can provide a strong foundation for school improvement. This process of assuring school improvement in low achieving schools includes:

- Identification
- Analysis/recommendations/planning
- Reorganization of programs and personnel, as appropriate
- Capacity building at the school
- Continued monitoring and adjusting until acceptable performance levels are reached.

The correlation research lays the groundwork for further, more experimental work to understand the specific practices actually providing the most leverage. For example, in a subsequent book, *Learning For All*, Lezotte discusses the original and second-generation correlates and identifies several "high yield" strategies for school improvement which are well researched by specific studies which control for other possible causes, and isolate interventions of proven leverage.

Yelon (1996) clearly pinpoints the danger of remaining at the observation/correlation level in telling the story of an observer of his wife's effective classroom. Fran Yelon had her kindergarten class pick up the classroom at the end of every day, while she played a song on a record. One teacher visited the classroom, but missed understanding the powerful classroom management practices Fran employed and simply said wistfully, "I wish I had that record!" We can do better, but we do better when we use systematic action research to tease out what really makes the difference and when we use quality improvement tools to identify and to continuously improve our core processes.

In the following discussion on addressing the goal of improving student achievement, we use the correlates of effective schools to explore these issues as well as strategies to move beyond learning from past and current success to moving toward what does not currently exist, to move beyond effectiveness to continuing excellence, and ways to move toward more sustainable improvement than what is typically experienced.

In this, the "heart of the matter" is the relationship and alignment between the high expectations represented by 1) the curriculum (the body of knowledge, skills, and character expected of all students), 2) the arrangements for learning and instruction (opportunity to learn), and 3) the assessment or continuous monitoring of learning. The motivation is, of course, high expectations for all. When these are well aligned and each is continuously improved sustainable transformation results.

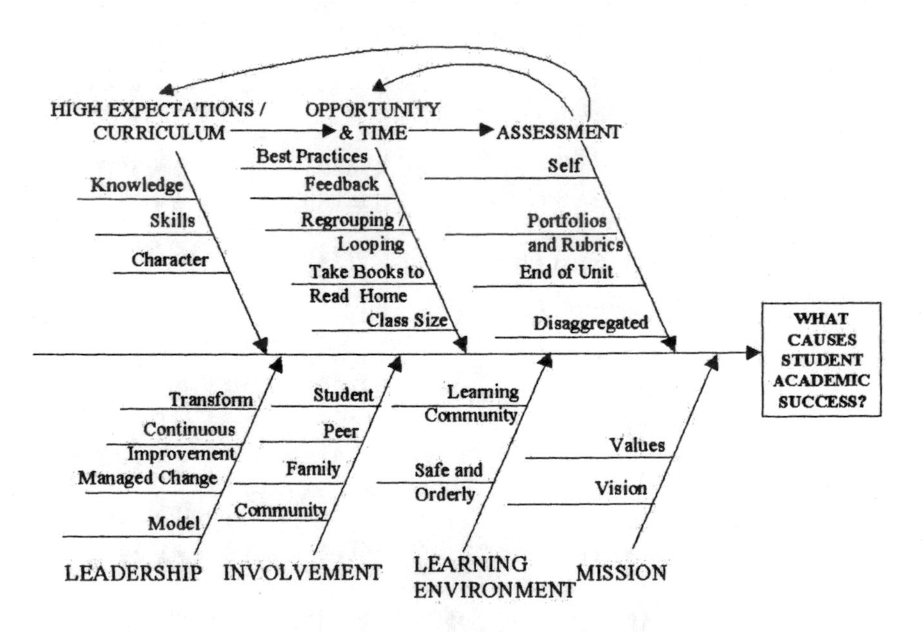

Figure 11.1. Clear and focused mission.

The transformation being advocated here involves nothing short of a complete reconsideration of what public schools in America really do, should do, and how they do it. Specifically, public schools may be only the latest in a long and distinguished list of enterprises, which nearly perished because they failed to focus on what their stakeholders/customers wanted or needed. As far back as 1972 Theodore Levitt pointed out the fate of businesses and whole industries that suffered from "marketing myopia." Railroads, not seeing that they were really in the transportation business, and Hollywood, defining themselves as in the "movie," not the entertainment business, nearly led to the demise of both.

Levitt (1995), describes several businesses, and says that to survive, they themselves will have to plot the obsolescence of what now produces their livelihood, mostly by defining their mission in the broadest possible terms and focusing solidly on their customers. The customer orientation is right out of the "total quality" movement, which brings with it a solid set of continuous improvement tools—ways to assess customer satisfaction, but also ways to gather improvement information to guide continuous improvement by first line staff. It also moves an organization to form partnerships with suppliers so as to build eventual quality for the customers.

All this learning from business and the use of business terms and tools has its cautions. Public schools serve human students and their parents. But applied with care, business concepts can and will help public schools to engage in sustainable development—continuous improvement—and assist us to not only address our mission more successfully, but will engage our public and each of our stakeholders in a way so as to build support at a time it is much needed.

More broadly, voices as diverse as Mortimer Adler and James Michener agree that both the mission and the methodologies of American public education urgently require sharpening, for the good of the schools, and for the good of the country. Michener sets a broad charge that is as timely as it is timeless when he proposes that public schools must lead the effort to advance democracy, into, during, and beyond the information revolution (Michener, 1996).

This linking of public education reform to the advancement of democratic ideals and the information age means actually modeling democratic practices and participating in the information age as well as teaching about them. Our students will go where we walk, not where we point. Therefore, the fundamental relationship of the individual and individual rights and responsibilities needs to be considered in relationship to the role of the community. We all want to be free and strengthen our independence, and sometimes at the same time we want very much to belong and strengthen our sense of security. Democratic ideals are so important, and so enduring to some extent, because they directly address the strength and the tension between individuals and the society or community within which they exist. Benjamin Barker and Patrick Watson (1988) note that, "The struggle for democracy turns out to be a struggle within ourselves: it is a tension . . . between the yearning for freedom and the yearning for security, the need to govern ourselves and the need to be taken care of by others, the need to give and the need to take"(p. xvi).

High Expectations—The Curriculum and Pupil Expectations

In considering continuous, sustainable school improvement, the heart of the matter is the careful articulation and relationships between the curriculum, the learning and instructional arrangements, and the gathering of feedback assessment. The driving consideration for these must always be high expectations for all. We must first be clear about the curriculum, then be sure we foster learning appropriate to *that* curriculum, assess learning and use that assessment data as feedback for continuous improvement.

As much as possible the curriculum should be a tangible manifestation of the driving goals of the school system and the total community. This can be accomplished through open meeting and surveys of community opinions.

Increasingly, it is clear that high expectations for all must involve all three dimensions—the knowledge, the skills, and the character—needed for individual success in sustainable communities.

Opportunity to Learn and Time on Task—Instruction

Finding out "what works and why" either by individual action research, organizational learning, or a review of the literature, represents the continuous adventure of the learning individual, study group, organization, or community. It means paying close attention to results, studying results, and especially arranging small experiments or action research to gather more and more meaningful feedback to guide one's own learning. All too often the answer to the question, "What works?" is "It depends." This makes the adventure a lifelong journey. Smaller class size, site-based management, block scheduling, increasing use of technology, all can get results—*depending* on exactly what happens within the practice. What goes on in a smaller class no doubt matters more than the number of pupils in and of itself. So, again, being clear about the purpose, and learning the exact relationships among the parts, is vital to careful continuous improvement. The continual identification, implementation, and refinement of "best practices" must itself become common practice. It is the investigation, the journey, which will drive continuous improvement, which will lead to sustainable improvement efforts.

Focusing for a moment on the top few practices, we come to reflect on some of the more powerful and long-standing answers to the question of "what works?" One-to-one tutorial settings generally lead to results on the magnitude of two standard deviations better than large group instruction. Apparently, that applies to peer tutoring as well as adult tutoring. Following that, reinforcement or feedback, continual monitoring of pupil progress and adjusting of the learning setting in response to the progress made, providing appropriate cues and explanations, participation and active involvement, time on task, and providing needed prerequisite reading and study skills, all have a significant impact on student achievement at the classroom level. (Bloom, 1984) Presumably, any practice, approach, or program which increases these conditions will do better than those initiatives which do not.

A recent, and very impressive report from the State Education and Environment Roundtable (Lieberman and Hoody, 1998), directly addresses the question of "What works?" The report, which gives a very commendable blend of standardized test results along with alternative, more authentic, measures of success, documents positive results of the common features of programs successfully using the environment as an integrating context for learning (EIC). These include:

- interdisciplinary integration of subject matter;
- collaborative instruction;
- emphasis on problem solving and projects;
- combinations of independent and cooperative learning; and
- learner-centered approaches—active participation.

Being clear about current classroom best practices raises the question of what are the "best practices" at the school level which foster effective classroom practices. We start with the continuing "effective schools research" around which this section of the discussion is structured.

1. Clear and Focused Mission
2. High Expectations/Curriculum
3. Opportunity to Learn/Instruction
4. Frequent Monitoring/Assessment
5. Supportive Learning Environment
6. Involvement
7. Leadership

But other sources as described below address the issue.

In addition to many state and school system practices where unsuccessful schools are identified, provided support, and encouraged to improve or face some sort of reconstitution of the school and/or staff, there is experience nationally.

Anne Lewis (1997), identified what she feels are features of successful schools:

1. A strong focus on academic success for all
2. No excuses
3. Experimentation
4. Inclusiveness
5. Sense of family
6. Collaboration and trust
7. Passion for learning and growing... Leadership of the Principal

Recent studies reinforce those ideas. It is best to start from the thoughts of someone who has done it—a turnaround principal. Dr. Linda Collins, Principal, Berry Elementary School in Nashville, Tennessee, and her staff responded to the challenge by moving average reading scores from the 24th to the 54th percentile. How? She says:

- Hard work
- Curriculum alignment
- Intervisitation between grade levels
- Careful implementation of instructional "best" practices—including frequent assessment of pupil progress
- Improved collaboration

Consistent with this example, researchers who visited sites as a part of a congressionally mandated study of Effective Schools Programs found that "the successful school-based reforms studied had three key features: 1) challenging learning experiences for all students; 2) a school culture that nurtured staff collaboration and participation in decision-making; and 3) meaningful opportunities for professional growth" (Educational Research Service, 1998, p. 103).

New technologies permit rapid "benchmarking," identification of similar but more successful schools, and help us capture exactly what practices might account for the success experienced in a given school. This will become a more interesting issue as we strive to identify new factors rather than continually reconfirming the known factors in new settings. The

Georgia Department of Education sponsors a unit specifically charged with this task; they have developed software needed to quickly identify similar schools across Georgia to visit and study for leads to greater effectiveness. This will require skills, not currently widespread, in first knowing what is leading to success and then in discerning the same from an outside perspective. All this can only be healthy for educators dedicated to continuous improvement, but it will involve specific staff development in analysis of successful settings.

And, let us not forget system level factors that lead to the opportunity to learn and to success. First, those factors that encourage effective schools and effective classrooms—staff hiring, evaluation, and development. Generally, it means inverting the pyramid to authentically provide support to schools, classrooms, students, and families. And money does make a difference. Harold Wenglinsky (1997) reports in his policy information perspective, *When Money Matters*, that "some traditional spending practices of school districts (spending for teacher-student ratios and central office administration) are conducive to academic achievement (p. vii).

Support through assuring appropriate funding has been found in another recent study to make a difference (Verstegen and King, 1998):

> . . . There are clear relationships between funding and achievement emerging from the recent body of production function research. These studies provide further evidence that money matters in producing educational outcomes. Given the close relationship between spending and both class sizes and teacher experience levels, this should come as no surprise, as these factors drive up the cost of instruction. However, it is clear that schools cannot be effective with resources they do not have (p. 262).

Frequent Monitoring of Pupil Progress—Assessment

Once the "what" it is we want students to know and to be able to do has been established and the opportunity to learn or effective instruction has been provided, it is essential that we determine if, in fact, learning has taken place and students have been successful in acquiring the skills, knowledge, and character deemed appropriate. To assure that success has taken place, a snapshot on a standardized assessment may be necessary but is not sufficient. Monitoring feedback throughout the learning process must take place in order to adjust instruction and opportunities to learn and to provide all the stakeholders—teachers, parents, and students—an opportunity for continuous learning and for taking responsibility for that continuous growth. If students do not know where they stand, they will not know what steps to take to continue. If teachers do not know how the students are progressing in learning, they will not know what steps to take to modify instruction and facilitate learning. Parents need to know how their children are progressing to facilitate and support learning for their children. As new initiatives are implemented, it is critical that the extent of pupil progress be assessed to determine if such initiatives should be continued, revised, or eliminated.

Frequent monitoring and feedback can and should be conducted in a variety of ways. These include classroom questioning and monitoring of responses, classroom assessments through a variety of techniques, such as standard multiple choice questions, open-ended questions, performance tasks, written essays, computer assisted assessment, journals, debates, projects and performances (individual or group), and portfolio development. Graded homework can provide additional feedback to students. These assessments can be used as a basis to determine if re-teaching is needed or if students are ready to advance to higher levels. A close match between the curriculum, instruction and assessment must be maintained—both in process and content. According to Smoker (1996), "Concentrating on results does not negate the importance of process. On the contrary, the two are interdependent: Results tell us which processes are most effective and to what extent and where processes need reexamining and adjusting"(p. 4).

A variety of assessment methods can assist teachers in monitoring progress. At the classroom level, teachers in many schools are assessing students upon completion of units of learning and frequently regrouping the students every two weeks according to their recent achievement levels. Not only are assessments conducted frequently, but also the results are used in making decisions on next steps in the teaching and learning process. Changes in instruction take place since the assumption is that not all students have learned what has just been taught. Teachers must be familiar with the curriculum, determine what is appropriate to assess, how best to assess if learning has taken place, and re-teach as needed. Teachers need to work together and learn from each other to plan units, assessments and instructional strategies. They also need to be assisted in finding methods to collect information and data in an efficient manner so as to provide the maximum amount of their time to instruction. The collaboration of teachers is critical to the process of regrouping based on achievement.

Other important classroom-level assessments include portfolio development—having students maintain work samples and reflect on their progress throughout the year. Computer-assisted assessments from a variety of educational software resources enhance frequent monitoring. Running records in reading can be maintained to assess progress.

The assessment rubric, part of a broader rubric relating to broad-based school improvement follows:

ASSESSMENT: FREQUENT MONITORING			
FOCUS	1	2	3
TIMING	Formal assessment is limited to end-of-lesson/unit activities.	The lesson-unit includes diagnostic as well as end-of-unit/lesson assessments.	The lesson/unit is assessed from beginning to end in ways that support and measure student learning, inform teaching and inform the learner.
ASSESSORS	Only the teacher evaluates students' work.	Student is asked to reflect in general; students interact with peers to share and give feedback but there is limited use of rubrics and checklists.	Assessments/artifacts include measures that guide student self-assessment and reflection on both products and processes (example, ongoing specific questions, checklists, rubrics); students may evaluate their own and each other's work.
ALIGNMENT	The assessments are unrelated to ESF curriculum; assessments do not measure student learning from the curriculum taught.z	Only some aspects of the ESF curriculum are measured.	The assessments/artifacts are derived from curriculum-embedded learning opportunities that measure and support learning.
EVIDENCE OF MASTERY	The assessment requires minimal response from student limited to answers to multiple choice, true false questions, or yes/no oral responses.	The assessment requires some verbal/written communication on the part of the student. This communication is limited to short test answers or question based oral responses.	The assessment requires an elaborate response of both knowledge/skills gained and process. This communication is provided through written, artistic, oral performances, exhibitions, artifacts and/or opportunities for students to teach.
AUDIENCE AND PURPOSE	The teacher is the only audience and the purpose of assessments is to test for a grade.	The purpose of the assessment/development of artifacts is vague or only school related.	The assessment/sharing of artifacts asks students to work for a real audience and purpose so that they can experience the benefits and consequences of their work.
ONGOING FEEDBACK	Feedback on artifacts is very general or ambiguous and given after the assessment is completed.	The assessment/ development of artifacts includes measures that provide specific feedback from the teacher after the assignment is completed. Revision is allowed but not encouraged.	The assessment includes measures that provide elaborate and specific feedback throught the process from both the teacher and peers. It includes measures that encourage all students to revise in order to produce quality work.

Figure 11.2

Systemwide assessments to provide feedback to teachers, students and parents are underway in many schools. The Georgia Kindergarten Assessment Program allows teachers to assess the students throughout the year on performance tasks related to three domains—literacy, mathematics, and social/emotional development. In Georgia, performance-based assessments are also administered in writing in grades 3, 5, 8, and 11.

Standardized testing at regular intervals allows for determining year-to-year gains for students. Looking at system and school averages on standardized tests tells us little of the progress of students. Achievement information on each student and the gains for each student from the previous year, however, provides extensive information. Instructional decisions can then be made not only on where the student currently stands, but also on how the student has been progressing. Teachers can reflect on what instructional strategies had the most impact on student achievement.

High school students are required to take tests in many states and are required to pass certain subject area tests at an established standard in order to receive a regular high school diploma. The results identify areas of strengths and weakness in the curriculum. Frequent monitoring and feedback must take place prior to that time to assist students and teachers. End-of-course core challenge tests help identify strengths and weaknesses early and to answer the question: "Are students acquiring the core skills, knowledge, and character being taught?" A recent publication from the Southern Regional Education Board reported that taking the "right" academic courses is one area found to make a difference in "High Schools That Work." They recommend that state policy and leadership initiatives include developing rigorous end-of-course exams and that data be collected showing the relationship between the courses students take and their actual achievement (Bottoms, 1998).

Teacher collaboration strengthens assessment. Teachers review the curriculum, develop assessment objectives based on the curriculum, and, develop items and performance tasks to assess those objectives. This collaboration between teachers provides a strong basis for continuous improvement. Item banks for use throughout the semester assists teachers in frequently monitoring progress and adjusting teaching as needed. These assessments can provide a closer match between achieved knowledge and grades.

More attention must be paid to individuals and significant subgroups. Frequent monitoring of the progress of every single individual student, and more disaggregated data to insure that gender, ethnic, and socioeconomic differences in achievement are not hidden within higher average scores and let stand is critical to focused improvement.

While opportunities to be authentically challenged need to be strengthened for *all students* through continuous improvement, most strategic planning efforts make special provision for currently low achieving students. When such students are found in high concentrations, "whole school" comprehensive improvement is called for. Careful identification of a realistic and appropriate standard is an important first step. Getting all school average up to the 40th percentile is an example. Self-study, augmented by expert teams, can and must lead to setting of priorities as to what high yield strategy ought to be addressed first. Extra developmental support in terms of resources is usually appropriate for two or more years. Continued monitoring and continual adjusting (re-organization), up to and including total reconstitution, may also be appropriate. This process of frequent monitoring of pupil achievement and making needed adjustments needs to go on until the standards are exceeded and the culture of continuous improvement is ingrained sufficiently to give a reasonable assurance of its sustainability.

Using the tools of continuous improvement such as benchmarking and control charts help provide a clear picture of how students and schools are functioning. Looking at control charts it is easy to identify schools falling below a specified benchmark, such as the fortieth percentile for two years in a row. Looking at percentile ranks that provide information about how schools and students are performing in relation to others is helpful, but most importantly looking at gain scores, helps provide further insight to a school's achievement. With expected gains at least 1.0 per year, it is easy to see where it may not be happening.

But, low achieving students are found even in high achieving schools and currently high achieving students can and must do better. Here the process of frequent monitoring and the making of adjustments would also occur, but would be sensitive to the schools current strengths so as to build carefully on them while making all needed modifications for the low achieving and all other students. In this fashion, "learning for all," is addressed both at the school and the individual student level. Other control charts graph cognitive abilities achievement levels. Some schools are achieving below the ability levels of students. Another way to look at the data is to determine the percent of students below the fiftieth percentile at particular grade levels with the lofty goal of moving all students at that grade level above the fiftieth percentile. In addition, progress is being made to measure students' enjoyment of reading as well as achievement. This is important since we know it is possible to teach a child to read but to not enjoy reading and that would be a big mistake.

Importantly, average and currently high achieving students are simply not challenged commensurate with their potential. Disaggregated data will help benchmark appropriate challenges, for every single student, and do so frequently during the year.

This means a new and expanded role for data, data formatted and presented as useful information. More data for its own sake may actually hurt improvement. More true information, which reduces uncertainty, helps. Organizational and individual learning from "what works" and "best practices," guided by a clear vision, will provide useful information for transformation.

PARADIGM SHIFT #2—MAKE EFFECTIVE AND EFFICIENT USE OF RESOURCES THROUGH AN ENVIRONMENT OF CONTINUOUS IMPROVEMENT

The shift is from reacting to external change and fixing things only when they are obviously broken, to continuous improvement or internal transformation, and then on to actually improve at the act of improvement. It is also a shift away from the selection and installation of "canned" programs, to the continuous movement toward fundamental principles governing best practices. This is less comfortable, but helps insure that an initiative and its underlying principles are understood. It means improvement and accountability based on improvements rather than successful installation of an "off the shelf" program.

Lezotte directs us toward the "principles of learning," Covey moves us toward fundamental principles of leadership and human behavior. In any event, the urging is the same, to continuously work toward practices based on fundamental laws and the measures of progress are necessarily customized to fit particular improvement initiatives. However, systematic application of respected criteria like the Baldridge Award criteria by internal participants and separately by external stakeholders will guide and give confidence to continuous improvement efforts.

In his entertaining and insightful book, *Who Moved My Cheese* (1998), Spencer Johnson reminds us that one of the best ways to minimize the stress of imposed change is to actually initiate change toward one's own worthy goals. He suggests we paint a realistic picture of "new cheese" and determine how to find it well before the current cheese runs out (or it gets moved by someone else).

Engaging in continuous improvement toward the future you prefer, before it is forced, is a time-honored and often ignored piece of advice. It was studied in depth by Ron Lippett and described in the book, *Creating the Future You Prefer*. This practice applies to individuals, organizations, and to entire enterprises like American public education.

Making effective and efficient use of resources is a concept close to the center of sustainable development generally and sustainable educational reform specifically. Clearly not a "one-time" effort, this brings us to a distinguishing characteristic of a learning organization—continuous improvement through continuous learning.

Creating the "learning community" is the key. Continuous improvement through continuous learning, for everyone—students and staff, only takes place in a conducive climate and we now know more than ever about that climate—that learning community.

Peter Senge (1994) describes the learning organization as one in which the following occurs:

> a. People feel they're doing something that matters—to them personally and to the larger world.
> b. Every individual in the organization is somehow stretching, growing, or enhancing his capacity to create.
> c. People are more intelligent together than they are apart. If you want something really creative done, you ask a team to do it—instead of sending one person off to do it on his or her own.
> d. The organization continually becomes more aware of its underlying knowledge base—particularly the store of tacit, unarticulated knowledge in the hearts and minds of employees.
> e. Visions of the direction of the enterprise emerge from all levels. The responsibility of top management is to manage the process whereby new emerging visions become share visions.
> f. Employees are invited to learn what is going on at every level of the organization, so they can understand how their actions influence others.
> g. People feel free to inquire about each other's (and their own) assumptions and biases. There are few (if any) sacred cows or undiscussable subjects.
> h. People treat each other as colleagues. There's a mutual respect and trust in the way they talk to each other, and work together, no matter what their positions may be.
> i. People feel free to try experiments, take risks, and openly assess the results. No one is killed for making a mistake. (p. 51).

A key reference in managing the changes that come with continuous improvement is *Managing at the Speed of Change* (1992) by Daryl Conner. Conner starts with being clear about the need for a change. He echoes the sentiments expressed earlier by Levitt, who said lack of a clear broad mission could lead to extinction. Conner intensifies that focus and describes the "burning platform" we must often visualize if we are to have the level of resolve Conner has observed in people who are successful in making significant change. He tells the story of a supervisor on a burning oil rig in the North Sea who jumped 15 stories into a freezing ocean where he could survive only about 20 minutes, and was already burning with floating oil and debris. "When asked why he took that potentially fatal leap, he did not hesitate. He said it was either jump or fry. He chose possible death over certain death" (p. 92).

Conner focus on "resiliency" and the character traits associated with it as essential for personal survival in a changing environment. He describes a resilient person as one who is positive, focused, flexible, organized, and proactive, and explicitly discusses the need for resiliency in the face of growing uncertainty and ambiguity.

Add impulse control and optimism to the growing list of highly desirable character traits emerging as needed in a complex, changing world and we see that managing change often means managing ourselves. While this will never be a popular message, those who learn it, and learn it early, as Daniel Goleman points out in his book *Emotional Intelligence*, will be rewarded for a lifetime.

Conner also is helpful in considering ways to organize for continuous improvement and the continuous changes involved in it. He identifies four key organizational roles in successfully managing change: 1) sponsors who have the power to sanction change; 2) agents who are responsible for managing the change; 3) targets who must actually change; and 4) advocates who want the change but lack the power to sanction it. As Conner points out, people often play more than one role. A person might be an agent for their boss, but be the sponsor to the people reporting to him.

Helping each role to be done well brings us back to the systems analysis. What matters as much as the roles being played well is the relationship between the roles. In fact, Conner specifically mentions the need for differentiating between initiating sponsors, who are usually high in the organization, and sustaining sponsors who are not as high, but have the relationship with the targets to persist with the commitment needed to sustain the change process. The sustaining sponsors have control of the more immediate consequences, and any time there is a gap between the strategic rhetoric and the local consequences, targets will be more responsive to the local consequences.

Consequences bring us back to the "burning platform" and another key reference in managing change. In the book, *Why Change Doesn't Work* (1996), authors Robbins and Findley describe the need to avoid either "pummeling" or "pampering" employees, and to keep the climate in a more moderate range they call "push" and "pull." The best is to earn the trust of the staff and realize the benefit of empowered employees pulling the organization to where everyone sees is a better place. A good example is the difference between a boss saying "improve quality or you are fired" (pummel or push) and an employee saying "is there anything in my work that could be arranged more efficiently?" (pull). The authors see pull as derived from Victor Frankl's book, *Man's Search for Meaning* (1997), and explicitly encourage recognition that we change by envisioning very intensely what we want to happen in the future.

While improvement, like all change, seems to necessarily bring some pain, we do have an obligation to minimize the pain as much as possible within our mission. And, much is known about change and continuous improvement.

It would seem that the school system goal should be to create the climate where all schools want to become "Learning for All" schools which will:

- Exemplify a desire to improve and to engage in authentic self-assessment.
- Identify, develop, and implement a well-defined School Improvement Plan focused on learning for all.
- Become knowledgeable of, implement, reflect on and document effective practices and programs that promote learning for all.
- Participate in action research projects to inform future continuous improvement.
- Share, adapt, and emulate the best practices known—based on own experiences and knowledge integrated with that of effective school research and others—to sustain continuous improvement.
- Participate in action research and study groups focused on effective practices.
- Use technology for analysis of data, turning data into useful information, and user-friendly storage and presentation of results.
- Become knowledgeable about research-based strategies, action research, learning organizations, and use of data to make informed decisions.
- Create reflective portfolios, which emphasize the learners (staff and students) taking responsibility for self-assessment and future learning and future improvements.

Peter Senge (1999) notes some of the problems associated with sustaining change:

> Fear and anxiety is included in the challenge of sustaining change because it develops after some progress is made in developing learning capabilities . . . effective leaders learn to recognize growing fear and anxiety as indicators of progress, and they learn how to acknowledge and deal with it, both within themselves and within others.
>
> This is difficult because it manifests itself in such diverse ways—as outright negation, as objection to ambiguity, as superficial support, and as silence . . . therefore leaders must learn to listen to the meaning behind the words, and find ways to reassure others that it might be natural to begin to see more significant, more difficult problems as progress is made on the most obvious problems. (pp. 242–243)

In order for continuous learning to take place, the climate must be conducive to that end. To improve effectiveness and efficiency, a learning climate must be established to create a learning community as described above. Another effective school correlate relates in this area.

Continuous improvement applies to efficiency as well as effectiveness. Using continuous improvement tools to reduce costs can be almost as meaningful as increasing effectiveness. Indeed, sustainability will at some level depend on schools doing better with less. Here benchmarking can be useful to compare costs to set realistic targets. Benchmarks help establish the gap to the best performance by a similar school or system, to a top performing school or system, and to a theoretically possible target. Benchmarks must be identified as target in critical areas for continuous improvement. Total quality management tools are available to measure progress toward the targets.

Widespread participation in painting a realistic picture of the future we prefer is a vital and often neglected step in managing change. Many stakeholders are addressing this by using a future graduating class as a focal point. Asking all stakeholders, especially current students, to identify knowledge, skills, and character traits that future graduates will need to be successful as citizens in the workplace and as individuals and family members can yield the picture that can energize change leading to sustainable improvement. As mentioned earlier, Alvin and Heidi Toffler have advanced one of the clearest, most universally respected views of the confluence between the information age and the needs of democratic ideals in the future. Kako (1998) has given an enlightening, if troublesome, view of technology from now to 2020, from 2020 to 2050, and beyond 2050. The graduates of the year 2020 will start to be born soon so our task is immediate. They will have the opportunities we want for them and will be more like the people we hope them to be if we plan now in specific terms for educating them.

PARADIGM SHIFT #3—IMPROVE STAKEHOLDER SATISFACTION, EMPHASING COMMUNICATION, RESPONSIVENESS, INVOLVEMENT, AND ACCOUNTABILITY

The shift is from the real and perceived isolation and unresponsiveness of public education to partnerships and meaningful accountability measures that actually encourage success for all students. Measures of progress will include data on student results, but will also include customer, stakeholder, and partner satisfaction surveys and public opinion polls.

The long and increasing press for accountability is the formal acknowledgement that schools belong to the community, not to the school staff. As simple as this sounds, the lack of acknowledgement of this truth looms as a major barrier to improvement, and to community support for the public schools.

What is needed is a thoughtful system of mutual responsiveness or mutual accountability among *all* the participants so that the system, its parts, and the relationships among its parts are orchestrated so as to meet the mission.

Specifically what is needed is a mutually interlocking set of expectations. The schools must seek participation and be responsive to students, families, and the wider community, and, the schools must be able to hold students, families and the wider community more accountable for the roles each must play in a successful system. In fact, all the parts need to be responded to and be held accountable.

Each level needs to have the flexibility required to not only get results, but to continually build capacity, to continually improve. In addition, each level needs to demonstrate the accountability required to appropriately support the other levels. This balance of flexibility and accountability will not ever be easy, or ever be settled once and for all. It comes fleetingly if, at all, as the temporary result of extensive teamwork and common commitment to a larger goal.

Indeed, as we explore what an "inverted pyramid" might be like, we experiment with the notion of treating other levels as "customers." This has meant moving on what principals and teachers have said they needed to be more effective, and has resulted in more flexibility and support at the school level. Principals have received purchase cards and higher limits on getting bids, more authority on who teaches in the school, and more time for school-level staff planning and staff development. We have also piloted "satisfaction surveys," filled out by central and building staff on top administrators, by principals on their supervisors, by teachers on principals, and by students and parents on principals and teachers.

Involvement

The key parts of the system under consideration are the student, the family, the classroom and the teacher, the school and its leadership, the school system, the school board, and the wider community. Each will be considered individually as follows:

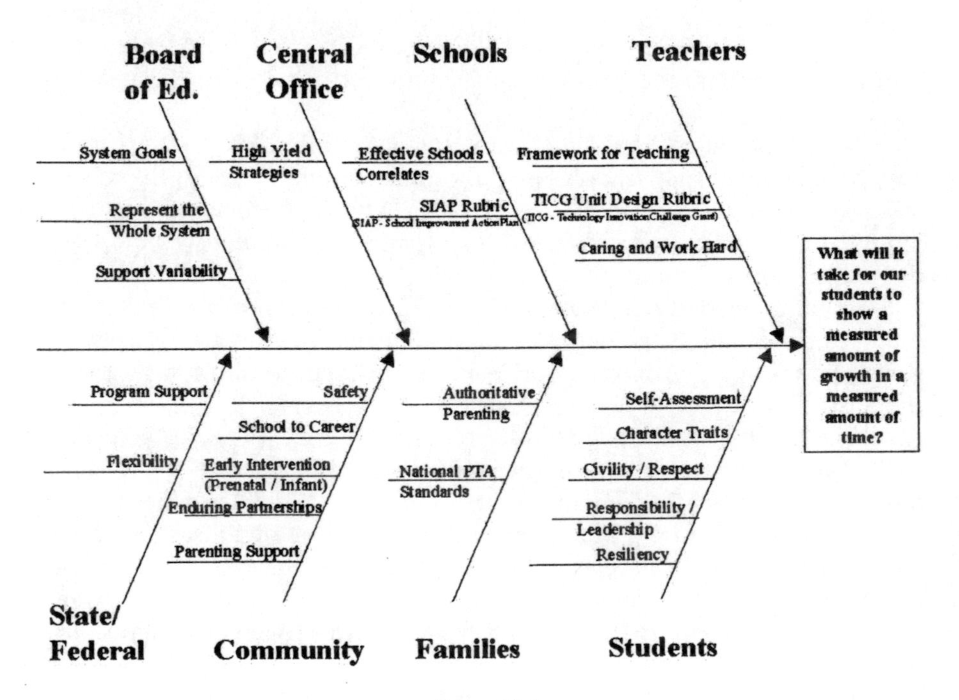

Figure 11.3

The Individual Student. It starts here, or rather it ought to start here, and all too often educational reform fads don't start with or ever even get to the student. The assumptions we make about students and role they play in their own education, probably makes all the difference in future school effectiveness. The relationships between students and the expectations among all the stakeholders matter a great deal, and are of little concern in many school "reform" movements.

Our metaphor or our mental model of the student matters greatly. In paraphrasing Plutarch, Rabbi Harold Kushner (1990) states his position in his remarkable book *When All You Have Ever Wanted Isn't Enough*, that people are more like plants to be nurtured than vessels to be filled with surplus wisdom.

Then, given the emerging vision of a "learning community," facing the implications *that we are all students, all the time* becomes most important. How do we want to be thought of in our role as students? How do we want to be treated?

Plants play an active role in their own growth. Certainly we can rely more on students to do so only if we change the way we approach them and the way we see our relationship with them. When we become partners we both learn, and we prepare the student and ourselves for the future.

Student portfolios, giving a comprehensive overview of the students goals and documenting progress toward those goals, open the door to useful self assessment and to students taking more responsibility for and involvement in their own learning. The portfolio is also a good place to summarize any contracts, compacts, or "stewardship agreements" made with other stakeholders. Importantly, the portfolio invites documentation of achievement progress, including but going well beyond, test scores.

A look at standardized test scores along with a more in depth look at student's work throughout the year will help identify those students who may need more time on task such as summer school. Those scoring below the 25th or 50th percentile might be identified for a closer look at their achievement throughout the year to determine if additional help is needed.

Without student active involvement in personal goal setting, progress monitoring, conscious adjustments based on feedback, and careful development of partners, which the portfolio encourages, the other dimensions of school improvement will be far less effective. Sustainable school improvement comes with students learning to and actually taking responsibility for their own lifelong learning.

Student leadership is also vital. Assisting students to not only take responsibility for themselves but for the classrooms, the schools, and the community is a task that must come first and is never abandoned. Character education in the early years, in partnership with schools but with parents taking the lead, is not only possible, it is necessary.

The Family. Here we are primarily focused on the relationship of the adults with the child, and the roles and responsibilities of the adults and the child. Increasingly we must face the fact that these relationships will be worked out with or without good marriages.

The national PTA Standards for family involvement provide an excellent framework for improvement in this area. Specifically, they draw attention to some often-neglected areas. Direct academic assistance provided by parents, and direct support by parents of disciplinary steps needed to foster a safe learning environment in the classroom, and on the bus.

National PTA Standards for Parent/Family Involvement Programs

Standard 1: *Communicating*—Communication between home and school is regular, two-way, and meaningful.

Standard 2: *Parenting*—Parenting skills are promoted and supported.

Standard 3: *Student Learning*—Parents play an integral role in assisting student learning.

Standard 4: *Volunteering*—Parents are welcome in the school, and their support and assistance are sought.

Standard 5: *School Decision Making and Advocacy*—Parents are full partners in the decisions that affect children and families.

Standard 6: *Collaborating with Community*—Community resources are sued to strengthen schools, families, and student learning.

Similarly, in *Beyond the Classroom: Why School Reform Has Failed and What Parents Need To Do*, Laurence Steinberg gives very pointed advice on how to avoid the pitfalls of over authoritarian and permissive parenting practices and benefit from "authoritative parenting." Schools will need to help students think earlier about effective parenting practices if we are to avoid the growing numbers of inadequate parents. Important considerations here come from, *The War Against Parents: What We Can Do for America's Beleaguered Moms and Dads* by Sylvia Ann Hewlett and Cornel West, where they examine the cultural norms and official policies that mitigate against strong families.

The Classroom and the Teacher. So much of school success rests on the teacher and the classroom it would be impossible to over-emphasize how vital it is to arrange for participation of teachers. A clear goal must be established to "invert" the traditional pyramid so that the central office and principals understand their roles in providing support to those whom directly work with students. The teachers, too, must be held accountable for the achievement in their classrooms. Setting a standard of at least a year's gain for a year in school for all students focuses on continuous improvement. In analyzing this type of data, however, one must proceed with extreme caution since the heterogeneity and starting points of classes may not be the same.

The School. The school has been the object of so much research on improvement that it is not likely to be forgotten in any improvement effort.

Nor is the central and important role of the leadership of the school principal likely to be minimized. What must be considered is what a school might look like in the future, and how past research might or might not apply when so much learning will be taking place at remote locations—at home, in the workplace in apprenticeships, and in the community. Technology and the demands for hands-on, applied learning will draw students away from lectures and passive learning in a classroom to active involvement and cooperative learning. Accountability at the school level for achievement has been discussed earlier and control charts provided for analysis.

The Support Staff. Secretaries, custodians, maintenance, bus drivers, cafeteria workers, and other support staff must be seen for what they are, vital links in a chain that must be strong at every point if students are to be served well and if continuous improvement is to become the norm. Formal and informal methods of communication must be created and maintained, and ways for all staff to thoughtfully influence the system must be provided

The School Board. The school board, in this country, can and should play a more productive role than generally has been seen to date. Only a few communities seem able to support boards, which manage change toward learning for all. They face support for the status quo from those who benefit from it, and opposition or indifference to the very changes needed to extend success to those who do not benefit at present. What is needed is for the board to represent all the students of the system, to set an enduring and clear direction, and to foster the local flexibility needed for local participation, understanding, and flexibility needed to be effective toward the common goals.

The State. In this country, education is a state responsibility, and considerable variation in performance between the states has been permitted. This has meant most "centralized" reform of schools and funding has come by state initiative or, more often, state level litigation. Recently state finance reform to insure equity of funding across school systems in a state has dominated state government attention to education. Little if any attention has been given to the question beyond equity, that of adequacy of funding.

The Nation. The federal government has sponsored a variety of steps to improve education. The original purpose of Title I was actually to improve the *capacity* of a school to address needs. It has instead been largely consumed in directly meeting student needs. Recent "whole" school efforts may bring it back to its original purpose and help us learn to fish, rather than simply passing out fish to schools where needs continue to grow and change faster than the federal government can respond.

Sustainable improvement will require much better coordination of the roles among the different groups involved with these groups taking responsibility for their results—hence, accountability.

Public Engagement. The public support has eroded and will not be regained with measures less than a total transformation of the relationship to the stakeholders. In the meantime, special interest groups seek to hold schools hostage to obtain their specific desire as the key to regaining general support—be it expanded programs for their population, lower taxes, general or specific cuts, privatization, vouchers.

Focusing the transformation of public education in this country on the individual student and the new roles and responsibilities of the student, and on the family, and on the leadership of the family as the first teacher and as the most important partner, supplier, and customer, will lead the transformation to a level of robustness and sustainability not found in other "reform" orientations.

The efforts to obtain a learning community will take leadership—another effective schools correlate.

Leadership

The main point about leadership in the transformation is that it will most certainly be shared and widely dispersed among the many stakeholders discussed in the previous section. Leadership will be clearly distinguished from management, and perhaps replace it.

W. Edwards Deming (1995) says, "This book is for people who are living under the tyranny of the prevailing style of management (p. vi)." He also asserts that, "the job of a leader is to accomplish the transformation of (the) organization.... How may he accomplish transformation? First, he has theory. He understands why the transformation would bring gains to his organization and to all the people that his organization deals with. Second, he feels compelled to accomplish the transformation as an obligation to himself and to his organization. Third, he is a practical man. He has a plan, step by step." (p. 119). "And, later a (leader) understands and conveys to his people the meaning of a system. He explains the aims of the system. He teaches his people to understand how the work of the group supports these aims . . ." (p. 28).

And, according to Deming, "a leader . . . has three sources of power:

1. authority of office
2. knowledge

3. personality and persuasive power; tact—the ability to appreciate the delicacy of the situation and to do or say the kindest or most fitting thing. A successful manager of people develops numbers 2 and 3; he does not rely on number 1." (p. 129).

Sergiovani (1991) wrote about moral authority following a conference with school superintendents from across the nation:

> Moral authority is derived from the felt obligations and duties that teachers feel as a result of their connections to widely shared school community values, ideas, and ideals. When moral authority is in place, teachers respond to shared commitments and to the interdependence they feel with others by becoming self-managing. (p. 8).

When moral authority drives leadership practice, the superintendent is a leader of leaders, a follower of ideas, a minister of values, and a servant of followership. (p. 12).

This focus on a leader in the context of a system is critical to sustainable transformation. A system is a set of parts interacting to achieve a purpose. If there is no purpose, there is no system.

The leaders in the transformation of American Public Education will be more responsive to more stakeholders, more accountable for results, more flexible in pursuing results, less bureaucratic and will put less emphasis on process rules. The leader will create a blend of the best of a "bottom-up" approach with the best of a "top-down" approach.

The leader will create and maintain more partnerships and give careful consideration of the role of the community and the role of the schools in serving community interests. The leader will encourage more learning from successful organizations, including businesses, and more sustainability—more "built to last" thinking, for our schools, as well as for our students.

CONCLUSION

Our opening quote made the point that "growth should be driven by quality and tempered by the long-term interests of the people . . ." Managing change, insuring that inevitable change will actually add up to improvement, and that long-term interests will be served greatly depends on America's educational system preparing students to advance democratic ideals while at the same time preserving both a strong economy and the natural environment. To do so the schools must transform themselves.

Huge transformations in the service of democratic ideals are possible and have happened before. Peter Jennings captured much when he said in 1989:

> "The cold war was over. It seemed an almost casual resolution to a sequence of stunning events: 10 years in Poland, 10 months in Hungary, 10 weeks in East Germany, 10 days in Czechoslovakia. The power of people in one place accelerating the pace of change in the next. The Soviet Union itself yet to come. China still to come." (Jennings, Peter, TV Guide, 12/30/89)

Incredibly, the Soviet Union did come.

Public school transformation, the break toward a focus on results, toward sustainable continuous improvement through continuous learning, and toward accountability and customer satisfaction, can happen. Our national passion for the march of democracy and our journey into and through the information age require the transformation of American public schools.

One "best practice" of the next few years may well be the thinking that will need to be done to figure out what knowledge, skills, and character the graduate of the year 2020 will need. The students of the class of 2020 will be born soon. They will need better prepared parents, better prenatal care, better infant brain stimulation, better preschool experiences, better education (beyond schooling) and a more "seamless" transition to citizenship and work if they are to be the citizens needed for democracy and for the quality of life during and what follows the information age.

Csikszenthmihaly, (1990) notes that:

> —humanity has achieved incredible advances in the differentiation of consciousness . . . differentiation has produced science, technology, and the unprecedented power of mankind to build up and to destroy its environment.
>
> But the complexity consists of integration as well as differentiation. The task of the next decades and centuries is to realize this underdeveloped component of the mind. Just as we have learned to separate ourselves from each other and from the environment, we now need to learn how to reunite ourselves with entities around us without losing our hard won individuality. (p. 240)

Within the public schools of America we have seen nearly all that is needed for now. Many islands of excellence exist, and much planned improvement has been implemented. Perhaps one of the big changes needed is knitting the islands of success together to serve all students, and that awaits more systematic thinking and acting. Institutionalizing such continuous improvement in our schools is needed if the public schools are to play the role they need to play for our national ideals, our economy, and our environment to be sustained in the future.

REFERENCES

Adler, M.J. and Blackmon, J.A. (1988). *We hold these truths to be self-evident: Understanding the ideas & ideals of the constitution*. Tappan, NJ: Macmillan Publishing Company, Inc.

Barker, B. and Wilson, P. (1988). *The struggle for democracy*. Boston: Little, Brown Company.

Block, P. (1996). *Stewardship*. San Francisco: Berrett-Koehler Publishers.

Bloom, B. (1984). The 2-sigma problem: The search for methods of group instruction as effective as one-to-one tutoring. *Educational Researcher*, June/July, p. 4–16.

Bottoms, G. (1998). *Things that matter most in improving student learning*. Atlanta: Southern Regional Education Board.

Conner, D.R. (1992). *Managing at the speed of change—How resilient managers succeed and prosper where others fail*. New York: Villard.

Covey, S.R. (1994). *First things first*. New York: Simon & Schuster.

Covey, S.R. (1989). *The seven habits of highly effective people*. New York: Simon & Schuster, Inc.

Csikszenthmihaly, M. (1990). *Flow. The psychology of optimal experience*. New York: Harper & Row, Publishers.

Danielson, C. (1996). Enhancing professional practice: A framework for Teaching. Alexandria, VA: Association for Supervision and Curriculum Development.

Deming, W.E. (1995). *The new economics for industry, government, education*. Cambridge: Massachusetts Institute of Technology, Center for Advanced Engineering Study.

Editorial. (1998, July 28). *Fort Myers News*.

Educational Research Service (1998). *Comprehensive models for school: Finding the right match and making it work*. Arlington, VA: Educational Research Service.

Frankl, V. (1997). *Man's search for meaning*. New York: Pocket Books.

Gingrich, N. (1995). *To renew America*. New York: Harper Collins Publishers, Inc.

Goleman, D. (1997). *Emotional intelligence*. New York: Bantam Books.

Hewlett, Sylvia Ann and West, Cornell (1998). *The war against parents: What we can do for America's beleaguered moms and dads*. New York: Houghton Mifflin.

Hunter, M. (1982). *Mastery Teaching*. El Segundo, CA: Tip Publications.

Johnson, S. (1998). *Who moved my cheese?* New York: G.P. Putnam's Sons.

Kako, M. (1998). *Visions: How science will revolutionize the 21st century*. New York: Doubleday.

Kushner, J. (1990). *When all you ever wanted isn't enough*. New York: Pocket Books.

Levitt, T. (1990). *Thinking about management*. New York: The Free Press.

Lewis, A. (1997). Washington commentary. *Phi Delta Kappan, 78* (10), 735.

Lezotte, L.W. (1997). *Learning for all*. Okemos, MI: Effective Schools Products, Ltd.

Lezotte, L.W. and Jacoby, B.C. (1992). *Sustainable school reform*. Okemas, MI: Effective Schools Products, Ltd.

Lieberman, G.A. and Hoody, L.L. (1998). *Closing the achiement gap—Using the environment as an integrating context for learning—Results of a nationwide study of the State Education and Environment Roundtable*. Poway, CA: Science Wizards.

Lorenz, K. (1963). *On aggression*. New York: MJF Books.

McKibben, B. (1998, May). A special moment in history. *Atlantic Monthly, 281* (5), p. 55–78.

Michener, J.A. (1996). *This Noble Land*. New York: Random House, Inc.

National School Safety Center. (1999). *National school safety center overview*. Westlake Village, CA: author.

Peters, T.J. and Waterman, R.H. (1982). *In search of excellence—Lessons from America's best-run companies*. New York: Harper & Row, Publishers.

Ravitch, P. (1985). *The schools we deserve: Reflections on the educational crises of our times.* New York: Basic Books, Inc.

Robbins, H. and Finley, M. (1996). *Why change doesn't work.* Princeton: Peterson's.

Senge, P.M., Kleiner, A., Roberts, C., Ross, R.B., and Smith, B.J. (1994). *The fifth discipline fieldbook—Strategies and tools for building a learning organization.* New York: Doubleday.

Senge, P., Kleiner, A., Roberts, C., Ross, R., Roth, G., and Smith, B. (1999). *The dance of change: The challenges of sustaining momentum in learning organizations.* New York: Currency Doubleday.

Sergiovanni, T.J. (1991). *The wingspread superintendents: A new definition of leadership.* Unpublished manuscript, Trinity University, San Antonio.

Strong, R., Silver, H., and Perini, M. (1999, March). *Keeping it simple and deep.* Educational Leadership, p. 22.

Smoker, M. (1996). *Results—the key to continuous school improvement.* Alexandria, VA: Association for Supervision and Curriculum Development.

Steinberg, Laurence. (1996). *Beyond the classroom: Why school reform has failed and what parents need to do.* New York: Simon and Schuster.

Toffler, A. (1971). *Future shock.* Madison: Demco Media.

Toffler, A. (1980). *The third wave.* New York: Bantam Books.

Toffler, A. and Toffler, H. (1995). *Creating a new civilization.* Atlanta: Turner Publishing.

U.S. Bureau of Education, Bulletin No. 35 (1918). Cardinal Principles of Secondary Education: A Report on the Reorganization of Secondary Education, Appointed by the National Education Association. In D. Calhoun (Ed.), *The educating of Americans: a documentary history* (p. 486). Boston: Houghton Mifflin Company.

Vergenstein, D.A. and King, R.A. (1998). The relationship between school spending and student achievement: A review and analysis of 35 years of production function research. *Journal of Education Finance, 24,* 243–262.

Wenglinsky, J. (1997). *When money matters: How educational expenditures improve student performance and how they don't.* Princeton: Educational Testing Service.

Yelon, S.L. (1996). *Powerful principles of instruction.* New York: Longman Publishers USA.

CHAPTER 12

Learning Communities
Vehicles for Learning Community and Sustainability

Jean MacGregor

"Far deeper than any question of curriculum or teaching method is the problem of restoring the courage of Americans for facing the essential issues of life. How can it be brought about that the teachers in our colleges and universities see themselves not only as the servants of scholarship, but also in a far deeper sense, as the creators of the national intelligence? If they lose courage in that endeavor, in whom may we expect to find it? Intelligence, wisdom, sensitiveness, generosity—these are not the byproducts of the scholarly pursuits—they are the ends in which all our scholarship and teaching serve."

Alexander Meiklejohn[1]

Seventy years ago educational pioneer Alexander Meiklejohn not only articulated this vision, he created an interdisciplinary college that immersed its learners in intense dialogue about democracy, community, and

[1] Meiklejohn, Alexander. 1932. Written assignment for sophomores reading *The Acquisitive Society* by R.H. Tawney. Quoted in Cynthia Stokes Brown, *Alexander Meiklejohn: Teacher of Freedom* (Berkeley, CA: Meiklejohn Civil Liberties Institute Studies in Law and Social Change No. 2, 1981) 99.

Jean MacGregor, Director, National Learning Communities Dissemination Project, The Washington Center for Improving the Quality of Undergraduate Education at The Evergreen State College, Olympia, WA.

commitment. Though short-lived, the Experimental College at the University of Wisconsin was the first spring in a long and gathering stream of curriculum reform efforts in higher education known as "learning communities." While learning communities vary in structure and theme, they share a common intention: to create deep, holistic learning experiences that foster connections among students, faculty members, and disciplines, and often, connections between college and community as well.

The contributors to this book are themselves calling for a new national intelligence regarding our earth and its systems, our communities and cultures, and the role we should and can play in healthfully sustaining all these processes. Furthermore, these proponents of sustainability education argue that new conceptions of curriculum and pedagogy are absolutely essential if we are to fully explore the complex interwoven threads of sustainability, and if we are to fully ground concepts about sustainability in actual community settings. Learning community approaches are a promising new conception. This chapter offers a rationale for restructuring the curriculum into learning communities, as well as descriptions of several learning community programs built around sustainability themes in three different undergraduate college environments: a large public suburban university, a community college, and a small liberal arts college.

LEARNING COMMUNITIES—A GROWING NATIONAL REFORM EFFORT

While numbers of educational innovations are currently called "learning communities" to emphasize their focus on collaboration and common purpose, the term "learning community" is being used here to refer to intentional curriculum restructuring efforts that remake the learning time and space for students and their teachers. Learning communities are "curricular structures that link or cluster courses, often around an interdisciplinary theme or question, and enroll a common cohort of students, so that students have opportunities for extended inquiry, deeper understanding, and more interaction with one another and their teachers as fellow participants in the learning enterprise." (Gabelnick, MacGregor, Matthews, and Smith, 1990) Two brief examples: at Bellevue Community College in Washington state, a freshman level English Composition course is linked to an American Studies course in "The Landscape" around the theme, "The Power of Place." At SUNY-Potsdam in upstate New York, "The Adirondacks" is a course cluster that includes coursework in Environmental Studies; Environmental Geology; Writing and Critical Thinking; Introduction to Ethics; and Introduction to Outdoor Recreation.

These intentional efforts to re-form the curriculum arose from the analysis that it is not merely the historic isolation of the disciplines and the increasing atomization of knowledge that inhibits exploration of the complex issues that face the earth and humankind. The problem also lies in our prevailing educational delivery structure in colleges: the course. The course, usually delivered in 50–75 minute fragments and usually competing for a students' time with one, two, three or even four other courses being taken simultaneously, is simply too short, too superficial, and too disembodied from other courses to have meaningful impact on a students' knowledge and understanding. For the undergraduate student, even a compelling course is often eclipsed by the larger unrelated collection of courses required in most general education programs.[2] Not only that, typical courses' short, rigid blocks of classroom time and the tyranny of broad subject coverage often limits the pedagogy to delivery-and-explication modes of instruction, rather than creating opportunities for extended applied work in community or field settings, or for reflection on the meaning of the coursework.

To address these issues, Meiklejohn's Experimental College swept away courses in favor of full-time study. Students and their faculty devoted two years to a dialogue about community and democracy; in the summer, the students conducted intensive studies of their hometowns. The "Ex College" (as it came to be called by dozens of loyal alumni) became a beacon for numbers of other interdisciplinary educational experiments that followed, most notably The Evergreen State College in Washington. Founded in 1970 as a public liberal arts college, most of Evergreen's curriculum is structured around full-time, team-taught programs of coordinated study. In the decades since Evergreen's founding, a growing number of course-linking and course-clustering adaptations have begun to spring up in all types of undergraduate institutions. Today, well over 200 campuses are offering learning community programs at all levels and in a wide variety of curricular contexts. Parenthetically, it is interesting to note that this innovation has evolved in parallel with the growing interest in integrated, thematic curricula in the K-12 sector.

Contemporary learning community leaders offer multiple goals for their programs: to create islands of coherence within otherwise fragmented general education programs; to increase student success and retention in the first year of college through a strong, community-building experience; to foster greater intellectual interaction between students and faculty; to create arenas for building interpersonal skills among students and for

[2] Learning community pioneer Joe Tussman makes this point compellingly in his book, *Experiment at Berkeley*, Oxford University Press, 1969.

enabling them to communicate across significant differences; to revitalize faculty members through opportunities for team-teaching; to create new curricular arrangements to address interdisciplinary themes; to structure larger blocks of classroom time for collaborative learning, field study, and community-based service learning; to stimulate students' sense of civic and social responsibility. The power of learning communities is that they are a relatively low-cost strategy for reaching toward many—or all—of these goals simultaneously.

These programs have widely variable structures, which run the gamut from very informal course links all the way to full-blown fully team-taught programs of interdisciplinary study. On some campuses, hundreds of students are enrolled in learning communities in any given year. Most are situated in the lower division, to provide community for freshman learners or a coherent pathway in the general education curriculum. Numbers of colleges have established learning communities specifically for under-prepared students, which often link critical reading and writing courses to larger lecture classes. Yet others have created learning communities to create integrated academic experiences in the major or minor. Assessment studies indicate that learning communities indeed live up to their intentions: these programs have been found to be powerful for student retention, motivation, achievement, and intellectual development. They also stretch and revitalize faculty members by involving them in stimulating opportunities for collaborative curriculum creation and teaching. Many programs have become powerful innovation sites on campuses, for new curriculum and alternative pedagogical strategies.

While numbers of campuses are regularly offering learning community programs built around environmental themes or regional studies, to date only a few have created programs explicitly about sustainability. The following examples represent different adaptations of learning community structures, moving from loosely linked classes that have a common project, to a 6-credit interdisciplinary course on sustainability, to a year-long exploration of the relationships between sustainability and community development.

INTRODUCING SUSTAINABILITY THROUGH THE DISCIPLINES: THE POSITIVE FUTURES INTERDISCIPLINARY PROJECT AT OAKLAND COMMUNITY COLLEGE

At Oakland Community College in Royal Oak, Michigan, three faculty members in the social sciences are using a sustainability theme to connect and apply the content of their three disciplines, while at the same time creating a space in each of their courses for students to envision a positive future for society. This miniature learning community program links three introductory classes in the social science: microeconomics (for which macroeconomics is a pre-requisite) taught by Ken Bratton, American government taught by Dennis Bartleman, and introductory psychology taught by Debra Rowe. The three classes, with their separate enrollments of about 35 students each, meet during the same clock hours once per week and an additional half-hour is added to the class schedule. All three classes come together in a 30–45 minute plenary session to work collaboratively on a creative futuring project which is carried out in teams of 8–10 students. Each student team has 2–3 students from each of the linked classes, so there is representation from economics, political science, and psychology on each team. The semester-long assignment is to create either a future scenario of a more humane society or a scenario of an environmentally sustainable society. Teams focus their work in multiple aspects of each of these broad topics, and then use the concepts from the three disciplines they represent to imagine what changes would have to occur in order for their imagined scenario to become a living reality. Using the concepts and reading from their classes, library and Internet resources, the students learn about many people proactively helping our communities, and gain a sense of the scope of problems and the resources required for solutions. The interdisciplinary student teams work to imagine a positive future, gather information and focus their scenario and their use of disciplinary concepts, and then present it both in written form and a collaborative presentation to the three classes at the end of the semester. The scenarios encompass a broad span of issues: family and community wellness, strategies for limitation of population and pollution, energy planning, strategies for alleviating malnutrition and hunger, more humane and empowering workplaces, to name just a few.[3]

The faculty work carefully to teach students, nearly all of whom are encountering an ambitious team project for the first time, skills in teamwork, brainstorming, consensus-building, and positive futuring. Toward the end of the semester, a "Volunteer and Career Fair" is held on campus. Leaders of social service, human rights and environmental organizations from the greater Detroit area speak with the students about their visions and their work, as well as about career and volunteer opportunities in their organizations. Students see that there are volunteers and professionals right

[3] Debra Rowe, Dennis Bartleman, Ken Bratton, Michael Khirallah, Martha Smydra, "Creating Positive Futurists/Citizens/Change Agents via a Learning Community Interdisciplinary Project." Ninth International Conference on College Teaching and Learning. Center for the Advancement of Teaching and Learning, Jacksonville Community College, Jacksonville, FL. April 1998.

in their communities reaching toward the very goals they are discussing in class. After the ten teams have shared their scenarios, a final exercise at the end of the semester asks each individual student to reflect on the commonalities and oppositions between the scenarios, the connections between environmental sustainability and creating a more humane society, and the steps he or she might take to move towards making a scenario a reality.

For the faculty team, the key outcomes for students in this project are both disciplinary and interdisciplinary thinking about real-world problems, the interpersonal skills necessary for making a positive difference in one's family, workplace and community, positive futuring skills, and most of all, an enhanced understanding of how their involvement in societal issues can make a significant difference. The psychologist on the team, Debra Rowe comments,

> "We designed this program to reduce cynicism and apathy in our students, and to enable them to see themselves as positive agents of change with the skills to be successful. This is not worthwhile only for the small groups of students we teach each year; it should be a core component of the general education requirements in the higher education curriculum. The skills that will make students positive change agents to make a society better are the same skills needed for the increasingly entrepreneurial economy they are about to enter. Our graduates need to be positive change agents for themselves, but the key question is, will they care about contributing to the larger good? If they don't develop a self-concept about their ability to solve societal problems and contribute toward creating a better society, they will live only on the level of personal gain. Higher education has the opportunity to help students develop a broader and more positive vision about their own relationship to society."[4]

It is interesting to note that the positive futuring project is not "preaching to the choir." For the most part, the themes of social justice and environmental sustainability are new ones for the students, in that when they register for these courses, few have noticed the fine print in the course schedule describing the interdisciplinary project. Yet, at the semester's end, most students indicate that the experience was positive, and their eyes were opened to the very serious nature of local, regional and global problems. What follows are some reflections by students, written at the close of the Winter 1998 semester:

> "This project was a very positive experience. I am glad I got to be part of it. It goes to show you that an education isn't just about showing up to class and getting a grade. It is about applying what you have learned toward your everyday life. I was made more aware that 'I' can actually make a difference in the world."[5]

[4] Debra Rowe, telephone conversation with author, Royal Oak, Michigan, August 7, 1998.

[5] Excerpts, end-of-semester written reflections by students in the 1998 Interdisciplinary Project, Oakland Community College, Royal Oak, Michigan. Printed with permission.

"I am a very hardheaded person, some might even say stubborn. I came to the understanding that the future we were looking at wasn't MY future. It was OUR future and it had to be something all of us on the team could live with. I think this project gave me a better understanding of the problems that people have in other parts of the world when trying to reach agreements. At the same time, it shows that the people trying to make changes are not necessarily interested in compromise."

"I think the most valuable gift that I have gained from this project is that it has made me think about all these issues. I have not really taken the time to research and wonder about what is going to happen to the environment until this project. Now I am even more conscientious about daily recycling and even try to promote it in my place of work. No one knows exactly what is going to happen in the future, but it will be interesting to see how many of the scenarios that were presented will come true."

"From this assignment I have taken with me the realization that positive changes come about only from positive attitudes and ideas. The scenarios presented really give me some wonderfully imaginative, positive futures. I feel it is imperative that we believe in these ideas. If we do not provoke some of the very changes in our scenarios, we might not be able to see a future any longer."

PRAGMATIC VISIONS IN THE FACE OF COMPLEXITY: THE STUDIES IN ALTERNATIVE FUTURE ENVIRONMENTS (SAFE) PROGRAM AT GEORGE MASON UNIVERSITY

An upper division examination of sustainability has recently taken shape at New Century College at George Mason, a large commuter university located in the growing suburban sprawl west of Washington, D.C. Founded in 1995, New Century is an ambitious interdisciplinary college whose curriculum is built around learning community structures and whose pedagogy is animated by active, collaborative and experiential learning with strong service components. Over eight hundred students are enrolled in this school whose leaders have embraced the task of shaping learning relevant for a new century. After moving through a general education sequence of learning communities in their first year, New Century students take half their subsequent coursework in New Century learning community programs, either team-taught or taught by a single faculty member. Studies in Alternative Future Environments (SAFE) is one such program. It serves advanced students. SAFE is the brainchild of Will Wattman, an educator with extensive experience in community planning and community development. When asked what stimulated him to conceive SAFE, Will reports that key seeds were planted when, in the 1960s he simultaneously encountered Buckminster Fuller, the brilliant inventor, global thinker and futurist, and Donella Meadows, the systems scientist and futurist most widely known for articulating the "limits to growth" argument. "Bucky said there is enough,"

Will observes, "and Donella said there isn't enough. This is the ground I've been walking with my students ever since, trying to find the harmony between these two perspectives."[6]

SAFE operates as a research enterprise with parallel activities in collaborative work and individual work. It begins in common reading on sustainability, particularly in opposing analyses of global problems that echo the Fuller/Meadows counterpoints—most recently reading the World Commission on Environment and Development's *Our Common Future* (1987) and well-argued critiques of it. The students also look at issues of consumption in the developed world, again reading opposing perspectives on the question, "Do we consume too much?" Then, the program moves to explorations of work-on-the-ground in arenas of planning and sustainability, both via the Internet and in surrounding communities. In the "studio" portion of the program, student teams examine a particular aspect of an alternative concept. Over the several offerings of the course, students have examined the possibility of sustainable alternatives within the university and have met with university officials to examine what is in place, and what future alternatives might look like—for residence halls, space and energy utilization and transportation. Students teams have also designed "The Bus," an alternative system of transportation and shelter, and have partnered with the Blandy Experimental Farm associated with the nearby Virginia State Arboretum to envision the components of a "researcher village" community. For the individual work in the program, Will Wattman encourages students to choose a topic of interest—food, energy, transportation, shelter are typical ones—and to research it with an eye to how emergent alternatives relate to their social and cultural contexts, and to what alternatives cost and how they are funded.

Electronic technology is key to the program. Students conduct most of their research on the World-Wide-Web. Wattman comments, "This field of sustainable futures is so alive and pregnant. Students sense their work is part of something that is international in scale and unfolding right now. The Web has an immediacy and an emergent quality you would never get through a textbook. And, students also discover they are not alone: there are dozens, perhaps hundreds of groups out there working on sustainability visions, issues and projects."[7] As well as using Web resources, students communicate with one another through "Town Hall," the university's Web Crossing Server, a software that enables synchronous and asynchronous discussions, and sharing of ideas and resources.

[6] Will Wattman, interview by author, Fairfax, Virginia, May 19, 1998.
[7] Will Wattman, telephone conversation with author, Fairfax, VA, June 5, 1998.

Wattman passionately believes that students must encounter the complexity and paradox of any examination of sustainable practices. For example, he says, "I often suggest that students read a new title that deals with sustainability from a feminist perspective. It begins with a wonderful, eye-opening anecdote by a writer who determines not to buy so much junk, while poring over her toaster in her too new kitchen and then realizes that the toaster was made in a Mexican sweatshop and what will the impact of her cutback have on the women who made it, etc. As the phrase goes, 'Gee this ain't easy McGee!' "[8] At the same time, he wants SAFE students to be "well armed and smart" about sustainable alternatives. "Positive visions," he argues, "have to be matched with pragmatic understandings of what is cost-effective, fundable, even third party fundable."[9]

In the SAFE program, the community of learning is built around the student team "studio" work, as well as continual sharing of ideas and resources in the individual student projects. Yet, students are also challenged to do the highly personal work of bringing the implications of sustainability into the realm of their personal lives. To that end, Wattman asks students to develop a "critical path," a construction of a map of personal expectations, wants and needs for leading a successful life. As the semester progresses and sustainability issues made more complex, each student is encouraged to revisit and revise his or her "critical path" map. Wattman comments, "The students are looking into huge problems: energy use on a national scale, or multi-housing systems, or alternative agriculture, but inevitably their work becomes reductionist. It inevitably puts *them* (that is, I, me) in the picture. The very large themes of the course ultimately depend on *my* individual behavior, on *my* response as a contributor to problems, on *my* acceptance of the problems and *my* willingness to move on them. So I want students to bring this home, to consider how what I have learned informs the rest of my life."[10]

LINKING SUSTAINABILITY TO COMMUNITY DEVELOPMENT AND SOCIAL JUSTICE: THE COMMUNITY DEVELOPMENT COORDINATED STUDIES PROGRAM AT THE EVERGREEN STATE COLLEGE

Sustainability concepts and themes have long been a staple of coordinated studies programs in environmental studies at The Evergreen State College. These programs, generally offered to upper division students, usually

[8] Wattman, personal correspondence, May 21, 1998.
[9] Wattman, telephone conversation, June 5, 1998.
[10] Wattman, telephone conversation, May 19, 1998.

comprise a year of full-time study taught by multidisciplinary faculty teams. Because students are enrolled in just one *program* rather than a constellation of *individual courses*, substantial blocks of time result, which allow for deep immersion in concepts, individual and collaborative research in the field or in local communities, and in applications projects in communities and organizations in southern Puget Sound. "Ecological Agriculture," for example, is a year-long program that provides a broad interdisciplinary examination of agriculture from the perspectives of "social, cultural and ecological sustainability."[11] Agricultural ecology, soil science, entomology are interwoven with history, policy studies, anthropology, and community studies to provide students with a comprehensive picture of the scientific, economic and political issues associated with sustainable agriculture. Another longstanding Evergreen program has been "Energy Systems," a year-long examination of scientific and social issues related to renewable and non-renewable energy sources, and issues and trade-offs associated with energy planning at local, regional and global levels. Other programs that run for one quarter explore sustainability concepts in the context of watershed and salmon resource management, forest management, and "sustainable development" in the third world.

Coordinated studies programs that focus on community development also appear frequently in the Evergreen curriculum; "Community Development: Conflicts and Strategies" will be described here because it represents a particularly successful effort to frame sustainability in community contexts, as well as to link environmental sustainability to issues of social equity and justice. In 1994–95, this year-long, full-time learning community involved 60 upper division students and three faculty members: Russ Fox, a land use planner with extensive experience in community development; Patrick Hill, a philosopher with strong interests in diversity and social justice; and Lin Nelson, a sociologist with expertise in environmental health and community-based organizations. Table 1 provides a program overview of the three quarters of study and community-based activity.

The faculty team agrees that it takes a year-long program to open the notions of community, community development, and sustainability and justice in all their complexity and to provide students the opportunity to identify, plan and carry out extended community-based research or projects. Shorter programs of a quarter in length can learn about communities, but year-long ones can learn in and with communities over extended time. When asked how sustainability was introduced in the program, Lin Nelson explained,

[11] Excerpt from "Ecological Agriculture" program description, The Evergreen State College catalogue for 1997–98.

Table 1. Overview of the "Community Development: Conflicts and Strategies" Coordinated Studies Program at The Evergreen State College

Quarter	Themes, student work	Readings, program activities
FALL	"Understanding the Issues and Conflicts" Exploration of theory and practice of community, social change and community development Examination of complex conditions: nature of democracy, participation, membership, inclusion and exclusion, and of broad structural forces: population, migration, globalization of the economy Student research and writing on "sung" and "unsung" heroes and heroines active in the work of sustainability Student portfolio of weekly integrative essays	*The Quickening of America, Rebuilding our Nation,* *Remaking our Lives*, by Lappe and DuBois *Community and the Politics of Place*, Kemmis *Beyond the Politics of Place: New Directions in* *Community Organizing in the 1990s*, Delgado *The Cultural Experience: Ethnography in Complex* *Society*, Spradley and McCurdy *Choas or Community? Seeking Solutions, not* *Scapegoats for Bad Economics*, Sklar *The Disuniting of America, Reflections on a Multicultural America*, Schlesinger *Theory and Practice of Environmental Justice,* *Toxic Struggles*, Hofrichter *Farewell to Matyora*, Rasputin *Vital Signs 1995*, Worldwatch Institute
WINTER	"Case Studies and Strategies of Community Work" Understanding and becoming involved in participatory processes which enhance community empowerment, racial and inter-cultural justice, environmental protection, and economic sustainability. Examination of case studies, dialogue with a range of community organizations, research papers and preparation for spring projects.	*Pedagogy of the Oppressed*, Freire *In Timber Country: Working People's Stories of* *Environmental Conflict and Urban Flight*, Brown *Race Matters*, West *Reclaiming Capital: Democratic Initiatives and* *Community Development*, Gunn and Gunn Case studies of the relocation of the town of North Bonneville, WA, the Grameen Bank and NGOs in Bangladesh, and the Mondragon Cooperatives in Spain. Field trip: Community toxics issues In-depth studies (students chose one or two)

Table 1. (*Cont.*)

Quarter	Themes, student work	Readings, program activities
		– Dialogues amongst conflict – Community planning – Community and environment
SPRING	Applying theories, perspectives and methodologies through community-based projects and internships. Poster sessions on collaborative community work; final student portfolios with synthesis essays and reflections on the year.	Weekly seminars to share experiences, resources, insights from community work: students chose one: – Environment and community – Youth, race and social justice – Urban and regional planning and agriculture Extensive end-of-year integrative discussions and writing on evolving notions of community, diversity of participation, the nature of collaboration and dialogue.

"It was absolutely critical for the students to develop an integrative framework as they examined theories of sustainability as they intersect community and justice questions. Rather than say, 'Here is sustainability and here are the multiple meanings of it," we embarked on a more collective sense of inquiry. We asked the students to identify a "sung hero" they thought represented sustainability, and to research that well-known person's work and present it to the class. Then, we asked students to write a biography of an "unsung hero," either a real person toiling away in anonymity or some imaginary person that would symbolize values and practices of sustainability. And through that, by their own discovery and learning, and teaching the rest of us about these people, students were engaged in their own emerging sense of what sustainability meant. We wanted our sense about it to be diverse and multiple. We wanted not to impose any particular definition, but rather to be critical of the idea and to understand how sustainability might not simply be defined, but how it is lived. We read the Worldwatch Institute book *Vital Signs* and took it apart, making presentations to each other on how the information in that book could be interpreted in multiple ways. Our intention in this was to stretch the contours a bit, to look at diverse perspectives on such issues as population and migration and development. Through these readings and others, films and the students already moving into some community ethnography work, we spent Fall Quarter trying to complicate the road map. There is really no one way to engage sustainability or community. We better be poised for complexity, poised for debate and controversy, while at the same time give these notions groundedness in people's real lives."[12]

[12] Lin Nelson, inteview by author, Olympia, WA, July 28, 1998.

To jump-start the process of contact-making by students for their community work in the Spring, Winter Quarter was launched with a resource fair of community organizations not unlike the one organized by the faculty team at Oakland Community College. Students were exposed to a rich array of community groups, many of whom identified specific projects with which they could use student assistance; the students would spend this quarter beginning their community research toward shaping a project or internship for Spring Quarter. Seminar discussions of the writing of Paulo Freire and Cornel West further developed concepts of community and justice. Two experiences particularly underscored these issues: reading *In Timber Country* (a book examining the impact of both environmental policies and regional demographic and economic changes on timber workers in the Northwest) and interacting with its author who came to campus as a guest speaker, and a day-long field trip to a working class community south of Olympia, where a low income housing community sits adjacent to a Superfund site of leaking PCBs and dioxin from a wood-treating facility.

> "Our visit with the community people in Chehalis coincided with a visit by the toxics activist Lois Gibbs, passing through on a speaking tour for her book, *Dying from Dioxin*. It was extremely moving to the students to go from readings on the science and policy issues related to Superfund sites, to listening to mothers worrying for their children's health, and to folks dealing with cancer clusters. From an environmentalist's perspective, this community could hardly be called an environmental hotbed. These were local folks, poor and disenfranchised, dealing with the complexities of Superfund, after years of toxic spills and controversy. Just being there for the day, meeting Lois and the townspeople and the local reporters was stunning for the students. The issue of sustainability got realigned for them around this issue, in that the local people were not using the nomenclature of sustainability, but they were struggling to find a way to sustain not only their health but the town's integrity. How do you find a way to talk honestly with your neighbors with this kind of fear hanging around? People were so nervous about even raising the issues, because of fear for their jobs."[13]

Another powerful aspect of the program was its linkage with the Washington State Corrections Center, located 25 miles north of the college in Shelton. Several prison inmates were enrolled in the program via technology (e-mail, and interactive television). Students enrolled on campus took turns traveling to the prison for weekly book seminars with their incarcerated fellow students, whose conceptions of justice and community, they discovered, were framed in a different reality.

These experiences with challenging reading, multiple perspectives, and complex community issues laid the foundation for students to solidify

[13] Lin Nelson, interview.

community projects to undertake in the Spring. The dozens of collaborative student projects and internships that resulted are too numerous to describe in detail here, but a few are worth mentioning. Some students particularly affected by their prison visits became active in prison education and issues of prison reform. Other teams became engaged in issues of troubled youth. Still others worked with environmental education, offering programs in local schools, establishing multi-generational community gardens, assisting with Olympia's annual "Procession of the Species" community celebration, and volunteering for the Olympia Sustainable Community Roundtable Initiative. A student who is a Laotian refugee worked with the state Department of Ecology to develop new ways for the department to notify and educate diverse communities, especially Asian immigrants, about toxics contamination in fish. Another group of highly committed wilderness activists in the program, who had started the year spending out-of-class time protesting logging practices in acts of civil disobediance found themselves meeting during in-class time with timber workers in coffee shops in logging-dependent towns on the Olympia peninsula. A sub-set of this group became so interested in the issues of sustainable futures for timber-depleted communities that they located an organization in Grays Harbor County, the Columbia Pacific Resource Conservation and Development project, a collaboration of former timber workers, tribal members and environmentalists attempting to assist troubled natural resource towns to find new ways to sustain their economies in a changing environment. The students offered themselves as interns to this group, moved to the community, and had an extraordinary experience doing research for the organization while engaging with dialogue with its constituents about the past, present and future of the region. Their work continued into the summer after the formal academic program's end.

As "Community Development" students' projects unfolded throughout Spring Quarter, they brought their experiences in communities and community-based organizations back to their faculty and fellow students, through the vehicle of weekly seminar discussions and through final project presentations at the end of the year. The year ended with a final exam in which individual students were asked to answer seven synthesis questions posed by the faculty, and three questions that each student individually posed for him or herself.

Looking back on the outcomes of the program for the students, Patrick Hill remarked that the experience "transformed students' polarized ways of looking at issues into a more multi-perspective approach, one grounded in more humble and collaborative and cooperative dialogue in communities."[14] Lin Nelson observed, "I took our challenge as a faculty

[14] Patrick Hill, interview by author, Olympia, WA, July 1, 1998.

team to support and enrich and protect students' deeply held passions for sustainability and environmental values, but at the same time to challenge them to think beyond them. This was not in any way to dilute or deflect their passions, but rather to risk looking at things in complicated ways that are way outside your comfort zone, and to reach out to people you disagree with, and find some common ground with them, while at the same time holding fast to your values. Sustainability can only become manifest through a lived experience, through real communities."[15] "For remarkable numbers of students in the program," Russ Fox added, "this changed their work and their life visions. To see themselves or their fellow students move from chaining themselves to trees to sitting down for coffee with timber workers, and then to putting on community forums to get people talking in ways they haven't talked before . . . or moving from going up to the prison to 'help the prisoners' to saying 'how do we educate people about the function of prisons in American society?' These students emerged with a whole different sense of identity of self and the communities around them. It was an incredible evolution. And, it could only happen in a year-long program."[16]

VISIONS

These learning communities are the forerunners of sustainability studies poised to emerge and grow in college and university settings. Programs like these offer both the structures and the pedagogy for bringing multiple disciplines to bear on sustainability issues, and for creating the time and space to move beyond the classroom so theoretical constructs can be tested in the lived realities of communities and work places.

Learning communities can deeply affect students when the notion of community is taken seriously—that is, when these programs intentionally attempt to build community among students and among students and their teachers, and better still, among students and community citizens or organizations. Collaborative academic work, particularly built around dialogue across differing perspectives, and careful attention to community-building can not only inform students' learning of content; it can also enable students to learn experientially the skills of community participation and community-making. Particularly when the community ethos fosters airing and listening with respect to differing viewpoints, students realize there are multiple stands on issues, multiple truths—*and that they have to make up their own minds.* When a pedagogy of diversity underpins notions of community

[15] Lin Nelson, interview.
[16] Russ Fox, interview by author, Olympia, WA, July 1, 1998.

in a learning community, a powerful environment for learning becomes possible.

Learning community programs can not only foster learning through community, they can set about learning *community* itself. By taking their studies into the real communities (both ecological and social) around our campuses, and by linking to community agencies and organizations that have substantive needs for project help, students can explore the multiple meanings of community, as well as apply their knowledge and skills in meaningful ways.

At this end of the century, there are echoes of Meiklejohn's ideals in the national conversations about the difficult challenges higher education faces, and in clarion calls for change. Arthur Levine and Jeannette Cureton's provocative book, *When Hope and Fear Collide: A Portrait of Today's College Student*, compellingly argues that colleges must strive to develop in young adults the attributes of hope, responsibility, appreciation of differences, and efficacy, the sense that one can make a difference. In fact, these authors argue that our institutions (our churches, schools, government and the media) are failing to provide our young people with these attributes because they themselves no longer have them. (Levine and Cureton, 1998, pp. 158–160) These attributes, these sensibilities, are foundational to "restoring the courage of Americans to face the essential issues of life" and to face the challenges of creating sustainable ecosystems and communities as well. As we set about the task of educating for sustainability, let us commit to nothing less than fostering and nourishing all these qualities—through learning *community*, and through learning *in* and *with* our communities.

REFERENCES

Gabelnick, F., MacGregor, J., Matthews, R. and Smith, B.L. (1990). *Learning Communities: Creating Connections Among Students, Faculty and Disciplines*. New Directions in Teaching and Learning Series No. 41. San Francisco: Jossey-Bass.

Levine, Arthur and Jeannette Cureton. (1998). *When Hope and Fear Collide: A Portrait of Today's College Student.* San Francisco: Jossey-Bass, Inc.

World Commission on Environment and Development (1987). *Our Common Future.* New York: Oxford University Press.

CHAPTER 13

Sustainable Development
Visioning and Planning

Alan Sandler

When you ask the question "what's wrong?" we found people just went to battle with each other. But when we ask "what kind of community do you want to be, not 2 years from now, but 20 years from now?" Imaginations creak open and conversations get on a different plane.

Jack Murrah
The Lyndhurst Foundation

SUSTAINABLE DEVELOPMENT

We all have a common objective to preserve and improve general quality of life over time. Sustainable development has proven to be a pliable catch phrase that planners, environmentalists, policymakers, and private developers have bandied about to describe/define this common objective. Sustainable development is essentially not so much about satisfying only environmental needs, growth needs, or human services needs as it is about satisfying all of these needs simultaneously and indefinitely.

The assortment of definitions for sustainable development reflects the diversity of issues falling under this concept, although a few fundamental

Alan Sandler, Executive Director, Architectural Foundation of San Francisco, San Francisco, CA.

ideas are consensually involved. Any community needs: a variety of social services to meet their needs, such as schools, hospitals, and police; a safe, healthy environment in which to live, with clean air, open spaces, and green areas; and economic vitality to provide for and support its residents. Problems must be properly addressed in the present using resources at a rate that keeps future use in mind. Individual localities should have a vehicle to envision their future as a unified community, a vehicle that helps communities reach those short- and long-term objectives of sustainable development. Such a tool can be a comprehensive planning and visioning process that commits to action.

Through an inclusionary planning process, communities themselves should identify the vision of what they want for their future, and advance the discourse on sustainable development into action. Sustainable development involves meeting present demands for economic growth in a context that provides for the long term safeguarding of social and environmental resources. Priority setting, resource allocation, and upholding civic interest are fundamental to achieving sustainable development.

PLANNING AND VISIONING

Planning for the sustainable development in the future of communities is an ongoing process of goal-setting and problem-solving that aims to bring about livable places. Planning focuses on ways of solving existing problems facing our communities, as well as providing a powerful tool people can use to achieve their vision for the future.

But what is visioning? Visioning is more than painting an idealistic picture of the future—it is a process of evaluating present conditions, identifying problem areas, and bringing about a community-wide consensus on how to overcome existing problems and manage change. By learning about its strengths and weaknesses, a community can decide what it wants to be, and then develop a plan that will guide decisions toward that vision.

Because change goes on all the time, a community must decide the specific criteria they will use to judge and manage that change. Instead of reacting after the fact to forces that alter their community, residents armed with a strategic vision can better reinforce the changes that they desire, and discourage changes that detract from the vision. Having a shared vision allows a community to focus its diverse energies and avoid conflicts in the present as well as the future.

A vision process must be participatory, without excluding existing groups or individuals who want to have a say in what direction the community should go, and how to get there. Participation makes for a better plan: the more people are involved, the more input; the more input, the better quality and quantity of information. And with better information, better decisions can be made about what can happen, and what needs to be done to make community plans actually come to life.

By including residents, businesses, nonprofits, local officials, and other community members in a process that empowers them as a whole, visioning also reinforces the idea that people do in fact have a say in their collective futures, and that they also have a responsibility to each other. Visioning is a process that all community members can have a stake in; the vision is their plan to provide for their future, not someone else's imposed idea of what it should be. For a visioning process to be effective, it must represent the people's vision, not only that of planners or bureaucrats.

Visioning is a practical method communities can readily use for their benefit. As a form of participatory planning, visioning not only allows for community members to interact, it also has the potential to open their eyes to problems and assets they might not have otherwise recognized. With more understanding of their strengths and weaknesses, communities are better able to seize the opportunities that will improve the quality of life for them and future generations.

A vision process may generally be comprised of the following steps:

1. Assess community characteristics, including information such as demographics, economics and employment, and housing. Identify community strengths and weaknesses. Have community members conduct a visual inventory on a map, indicating historic sites, scenic views, problem areas, and other notable places.
2. Identify the community's most important issues and questions.
3. Translate these issues and questions into goals and objectives.
4. Identify and discuss the obstacles to these goals and objectives, and the potential use of available resources to address them.
5. Conduct a "reality check," considering local attitudes about the goals and objectives. What kind of priority should the goals and objectives be given, considering the overall vision.
6. Discuss how the implementation of these goals and objectives will occur. Factors such as cost, time frame, and oversight are essential.
7. Decide how the vision's implementation will be evaluated, over what time intervals, using what criteria, and by whom.

VISIONING FOR THE FUTURE

In order to create a vision for the future, assumptions about what will occur are required. Assumptions are the underlying beliefs about existing or projected conditions that will impact on the community. Assumptions evolve from the data that has been gathered. They form the foundation for planning. A change in an assumption may necessitate a change of plans. Assumptions generated from the government planning process are not the only assumptions communities need to be aware of. Communities need to review their own data and incorporate assumptions from that data as well.

At this point in the planning process the emphasis is on:

1. engaging citizens in a process of:
   - reflecting on the history and present reality of the community(s);
   - understanding the recommendations and owning them as assumptions for planning for the future as a single community, a cluster of communities or as a new entity;
   - articulating a vision for the future;
2. assisting the leadership in:
   - clarifying assumptions from the data generated by the community(s);
   - validating or revising the mission and goals of an individual community and/or formulating a mission for a cluster of communities or a new entity;
   - formulating a vision and directional statements;
   - ritualizing and promulgating the vision for the future.

This is a creative time. All the data has been gathered and synthesized. Some of the data has been shared to arouse curiosity and challenge. Citizens want to see what comes next and they want to be involved in a meaningful way. The visioning process is key to the future of the community, the citizens are ready to build a shared vision.

SHARED VISION: the result of an interactive process in which members share their personal visions and shape them into a shared vision providing energy, coherence and direction for the communities' diverse programs and services. A shared vision answers the question,

"What do we want to create?"

VISION: is an idea or image of a desirable future which captures the commitment, energy and imagination of key people in working toward its realization.

THE RIGHT VISION:

- answers the question "What do we want to create?"
- reflects the personal visions of the people because it is rooted in the individual's sense of values, concerns and aspirations
- is appropriate for the community and for the times
- encourages commitment, inspires enthusiasm, and energizes people to act
- creates meaning in people's lives
- is well articulated and easily understood
- connects people together by common aspiration and provides focus and energy for learning
- establishes a standard of excellence and high ideals
- clarifies purpose and direction
- bridges the present and the future
- creates a sense of commonality which permeates the community and gives coherence to diverse activities

TIPS FOR VISION-BUILDING

The following are some tips to keep in mind in planning for the visioning process.

- The process of envisioning is as important as the outcome. Involve as many people in the process as possible. Keep the question of vision and purpose open and alive over time.
- During the creative phase, don't be constrained by present realities. Give yourself permission to explore, to play!
- Focus first and completely on the result of your effort, not yet on actions to achieve it. State the vision as if it is already true, in present tense and positive language.
- Expect the need for a quick, symbolic "sign" that the vision is more than lip service. Initiate the vision statement with some form of celebration that generates genuine energy and respect.
- The style and process used for announcing your vision must embody the spirit of the vision itself.
- Place someone in charge of keeping the vision alive, congruent and consistent in all aspects of the community. Constantly ask the question: "How does what we are doing support the vision?"

- The feelings evoked by the vision may threaten some people. Encourage open airing and testing. Periodically check "is this what we really want to create?"
- Remain flexible and adaptable to people's continuing input.
- The more people hold this vision actively, the greater its power to the change effort. Create a critical mass of support for it.
- Beyond talking the talk, you must walk the walk with full integrity. Ensure that top leadership embodies the vision in words and behavior. The community needs to live it, not just think it.

The visioning process is an opportunity to refocus the community's direction. Be creative! Do something entirely out of the ordinary in the initial design process. Take a risk in your plans for shaping this statement. Given its importance, take the time it requires. Use techniques that break the normal routines. Try using film, music, graphics, symbols that inspire creativity, guided visualization and right-brain activity.

This visioning process can be used in communities of any kind, i.e., classrooms, supervisory systems, towns or cities. Imagine the future if we all gained these skills, and were involved frequently in future visioning process in the key communities in our lives.

CHAPTER 14

Understanding Sustainable Communities

Larry Peterson

1. DEFINING "SUSTAINABLE" AND "COMMUNITIES"

Sustainability means, quite literally, the ability to keep going for a long time. The Japanese word for *sustainability* is *ji kyu ryoku*. These characters each mean possess, long, and power; or, "Possess power for a long time." As many definitions do, this one leads to more questions about what *sustainability* really means. What kind of "power?" How do you "possess" it? How long is "long?" And since this definition implies action over time, *what* is it you are trying to sustain? Sustainability is just as difficult to define as *healthy*.

Communities are almost as complicated as *sustainability* to define and certainly as difficult to understand. Coming from the Old French and Latin root words for *commune*, which means "common," communities includes things held in common, like government and social structure as well as a common sense of place or locale. Boundaries for these different elements of community may be difficult to draw; however, individuals within communities, whether fish, birds, or humans will know what community they belong to and where the important boundaries are. For most human individuals, some of these boundaries will change over their lifetimes.

Larry Peterson, Director, Florida Sustainable Communities Network, Florida A&M, Tallahassee, FL.

Children have different "home ranges" that increase in size as they grow older. As adults, humans today have access to mobility that would have been unthinkable just one hundred years ago. Until the invention of the telegraph, no human activity moved faster than a horse can travel. All goods, communications, and human travel moved by horse or horse-drawn wagons. The telegraph launched communications toward the speed of light and the steam locomotive moved goods and humans at speeds of thirty miles per hour. Both of these 19th century inventions dramatically changed and fractured our sense of community as a place-based social system.

Depending on how tightly knit the community is, individuals may not clearly know all the components of the community to which they belong. Wild animals obtain this knowledge through genetics, intuition, and early imprinting from their parents, peers, and their surrounding environment. They have no knowledge of what they have not seen or experienced. Their pattern of knowledge is constructed to deal with a range of environments (some are small and some are large) within which they can adapt and survive. Humans by contrast, gain knowledge of their community through similar means as wild animals, with the important addition of the education process. Humans purposefully educate their offspring with specific goals in mind. The immediate intention and hope is that this process will ensure the survival of the cultural system and perpetuate the ideals of the community. The longer term intention and hope is that the process will ensure the survival of the species.

Wild animals have little choice of both individual identities and communities to which they belong. There are some rare exceptions where wild animals raised in captivity will adapt and mimic the behavior of offspring from a different and locally dominant species. Absent individuals of their own species, some wild animals will adopt a community of another species and attempt to fit in. Humans, on the other hand, are very adaptable and can be programmed and re-programmed by their parents (which may change during the lives of many children), their environment (which certainly will change during their lives), their peers (which will also change), and their social and governmental structure (which is less likely to change).

Most importantly, humans enjoy more freedom of mobility than any other species on our planet; and they exercise this freedom more than any other species. Humans have developed very sophisticated methods of mobility and become attached to them. Exercising this freedom of mobility is threatening to the development of a sense of place and an understanding of ones' locale. The concept of topophilia, or love of place, is developed over a long period of time. Frequent movement fractures this development, undermines the development of community, and the parts do not necessarily add up to any meaningful whole. Unfortunately, this has not been a concern for many people enjoying the prosperity and freedom of the 20th century. As we enter the 21st century, this tide may be beginning to turn.

To be "placeless" is the definition of domesticated, or to be without an *intrinsic* place. Domesticated animals will move anywhere the owner takes them, and they will try to live there and stay with their source of sustenance. They are dependent on their owners (or animal "husbands") to provide for their survival and many domesticated animals will die if suddenly left completely on their own. Domesticated animals are bred for their ability to adapt to the environment the owner wants them to inhabit. They are adaptable and sustainable in this limited symbiotic relationship with their owner. Many wild animals, by contrast, are very attached to specific places and more limited in their mobility and adaptability. Some species will simply die if moved only a few miles from their home locale. Other migrating species move from specific place to specific place to acquire their life-cycle and seasonal needs. Humans have learned to take the components essential to maintain their chosen lifestyle with them when they relocate, even temporarily. If they cannot take the components they need for survival and comfort, they will take the tools necessary to make them. The range of "essential" items for humans has ranged from Winnebagos for many North Americans to a small leather pouch for Australian Aborigines.

Humans have not been around the earth for a very long time compared to some other species. Springing from a few locations, probably in and around North Africa, Homo Sapiens migrated in all directions to navigate the oceans and populate all of the continents. This process took place slowly, over hundreds of thousands of years, one mile at a time. Changing color, shape, and stature in the process, the variations in humans around the world today show the success of their adaptive capability. In a relatively short time, humans have learned to live in all climates on earth, created many variations in languages, art, and cultural artifacts; and developed religious, governmental, and social structures of very different types. All of these extensions, expressions, and fabrications of the human mind have the same eventual goal as the basic activities of other species of the plant and animal kingdoms—to perpetuate themselves through endurance, replication, or repetition.

Putting all of these concepts together provides a rough understanding of what *sustainable communities* means. *Sustainable communities* are collections of individuals that hold several important things in common: their sense of place or locality; their social, religious, and governance system; and they derive from both their individual interactions and their surrounding environment, the power to adapt to changing conditions and remain intact for multiple generations. Most individuals who belong to communities of this type are very loyal and respectful of their historical traditions and derive personal satisfaction and happiness from them.

Remarkable as human communities are, they do fall apart (or disintegrate) and sometimes come back completely changed; and they occasionally disappear altogether. Changing environmental, social, or governmental conditions can generate stresses that the community cannot bear and the only available survival strategy is to alter itself dramatically. Often the introduction of a new technology will trigger a chaotic leap in a community system and it will become quite different in a short time as it absorbs and diffuses the implications of the new technology and the new ideas that usually accompany it. Other components of the community will be stressed by this absorption but may survive relatively unchanged.

The real diverse wealth of a community can be assessed by its reaction to an important threatening change or disruption. The accumulation of personal or individual wealth is an attempt to sustain oneself or family and to provide security and reliability of comfort. The accumulation of community wealth is less tangible. It is certainly appreciation of the qualities of both the surrounding and immediate built and natural environment; and it is the integrity of its civic, social, religious, and governmental institutions. But, it is also the resilience and toughness of its residents and their ability to stick together, face difficult events, and emerge stronger from the experience. This attitude expressed by the community can clarify and strengthen values as well as reaffirm the common ground on which the community stands. The question, "What do you stand for?" has physical as well as ethical and moral dimensions. When a judge enters her/his courtroom and the bailiff orders all those present to rise, the community is re-affirming its pledge to stand for justice.

This important intangible element of human communities is the "glue" that is generated by the individuals in the community as they address threatening or stressful encounters. This intangible element can be quite strong and can bond individuals together as a community, or as part of a community in dispute with another part. The contentious drive towards consensus and unanimity, modified to "majority rules" in our government, is part of the basic glue of our society. Other, less dramatic activities also can create this bonding, like racial or religious affiliation, participation in a political party, or to be more contemporary and possibly trivial, liking a favorite sports team or the make of automobile you drive. The point is that all of these activities and choices create a sense of identity in common with others. Some are far stronger then others. Individuals will change sports teams when they move to another location; but probably not their religious affiliation as easily. Some individuals feel so strongly about specific issues, like abortion, that they will change political affiliation to support candidates who take their position. Even with extreme dislocation from their "home" place, humans will often continue to identify with their early imprinted locale, and continue to support its existence and traditions.

In the U. S., the overwhelming choices enjoyed by individuals in many of these important community factors have eroded the sense of community in many places. With our individual mobility and frequent change of family location, a deep sense of community is difficult to achieve and a sense of identification difficult to maintain. The homogenization of our suburbs and the deterioration of our urban centers have depleted the sense of belonging to some special place and the willingness to sustain it. In addition, relative to populations of humans in other countries, United States residents enjoy cheaper access to energy and natural resources and use them less efficiently than other cultures, with the resulting pollution of water, soils, and air. Exercising our freedom and mobility has its costs, both in environmental pollution of all types, and in loss of a sense of community.

The energy and resources of all types necessary to sustain our lifestyles and our culture is limited and the rate at which we are utilizing them is alarming. The current increasing attention being paid to *sustainable communities* is warranted and deserves our collective best efforts before our culture is overtaken by events that will trigger a chaotic leap into an unknown future. One characteristic of a purposeful human system is its desire for predictability. Our culture wants to predict, and hopefully control, its own foreseeable future. Even if we cannot control the events likely to evolve in the future, at the very least we would like to not be surprised by them.

2. A SUSTAINABLE COMMUNITY IS DEFINED BY SUCCESSFUL RELATIONSHIPS AMONG THREE INTERLOCKING SECTORS: ECONOMY, SOCIETY, AND ENVIRONMENT

Economy

Economics studies the process of placing value on resources in the marketplace. Resources of all types—human, natural, and capital—derive value from their interactions in the free market. In other words, in our free market capitalist economy these resources have no intrinsic *economic* value in themselves. They only derive economic value in the marketplace. For several centuries, this system of valuation has worked fairly well for human societies. Recently, we are beginning to see that the "free work of nature" has value for humans, even though it is not traded or accounted for in the marketplace. We are realizing that dollars do not pay back the earth for timber or the oceans for fish. The dollars only go to the humans and their companies that extract, harvest, and capture these resources and bring them into the marketplace where their price is determined.

Many ecologically minded economists have attempted to place an economic value on this free work done by natural systems on the behalf of humans. Various accounting systems have been generated to determine the marketplace "value" for the work forests perform as they cleanse our air of pollutants, and for the work swamps perform in cleansing our surface waters. Difficult as this task is and as unreliable as the final numbers are, the effort to define this free work is significant in its own right. It acknowledges that our free market system of economics does not account for everything necessary to sustain human life on this planet. While we will never "pay" the forests, oceans, and swamps for their work, we will gain an appreciation for them and have some comprehension of their worth to us as individuals and to our culture.

Another shortcoming of our economic system is its difficulty in assigning value to our resources held in common, or in the public interest. Splendid and magnificent physical resources, both natural and made by humans, are purchased by public sector agencies and literally taken off the market to be held in perpetuity away from market forces. The necessity for these actions alone should give us insight into the limitations and potential dangers of our market system to operate on behalf of our collective public interests. While we have governmental and non-governmental organizations using market forces and mechanisms on the public's behalf on a case-by-case basis, the marketplace itself has no inherent public interest. And yet, we continually and more predominantly view our lifestyles and culture through economic lenses and either directly or indirectly assign economic value to them.

The old saying, "You do not become conservative until you think you have something to conserve" is difficult to conceive in the free marketplace. Conservation of a resource does not necessarily create value, so the free market capitalist says, "Why bother?" Our economic system can only derive value for a resource when it is available for use. Consequently, concentrating on efficient use of resources is, in one sense, a de-valuing activity because it focuses on using or consuming less. At one level of economic analysis, efficiency makes sense if it increases profitability by lowering costs of resources in a given process. Reducing the costs of production while maintaining or increasing the price has always been a goal of manufacturing industries. However, if all the manufacturers in the U.S. become more efficient and use less petroleum, the petroleum industry will suffer financial shortfalls and have to raise prices and/or expand into new markets to maintain revenues and profits.

Placing value, even economic value, on public places and spaces for public activities is not strictly a marketplace activity. It is more a social and governmental activity for the greater benefit of the public. The public realm has value to society even if it will never command a high price in the marketplace. The community, through its sense of place and attachment to locale, will determine the value of its public realm resources and protect them from deterioration through collective activities. These activities may be political in nature or voluntary donations of money and time. Historically, providing for and protecting the common weal (or wealth) was the work and duty of all. Over the last centuries that role has increasingly devolved to the patronage and responsibility of wealthy individuals and large institutions. More recently, non-governmental organizations (NGO's) and non-profit organizations have solicited public memberships and obtained the financial and political means to act on the public's behalf.

Society

The society, taken as a whole, is relatively unpredictable in its specific behavior and fairly predictable in its patterns of behavior. The study of societal responses to issues of community sustainability reveals repetitious patterns of behavior. These patterns of behavior are directly related to the size of the community, as experienced by its collective individuals, and its relative position, physically and culturally, in the next larger aggregate of social responsibility. A small inner-city neighborhood threatened by an intrusion into its territory will follow a fairly predictable pattern of behaviors as it perceives and responds to this threat. The outcome of its efforts is also altogether too predictable as it collapses, gives in or gives up, or negotiates in support of its causes for concern. Small rural towns are just as predictable in their patterns of behavior when new development opportunities arrive to challenge and change their sense of community. There are exceptions, of course, but not enough to keep the tide of homogenization from rolling across the American landscape. From patterns of land development around rural turnpike intersections to patterns of urban flight to the suburbs and the gentrification of old urban neighborhoods, the patterns are monotonously clear to see.

A community's sustainability is measured by its ability to adapt to forces of change, both from within and without, and retain its identity and core values. Repeated challenges to a community can take their toll, with the community gaining clarity of its common ground and gaining strength, or with the community disintegrating and its values becoming more dilute as old members depart and/or new members arrive with different attitudes, values, and experiences. Our open, mobile, and relatively affluent society makes community-building difficult, if not impossible in the traditional sense.

The U.S. is entering a neo-nomadic, hunter-gatherer style of existence with individuals, either acting on their own behalf or for the benefit of their immediate families, migrating to more rewarding and profitable employment grounds when opportunities develop rather than staying in one locale in response to the more intangible values of place. Permanent residents of many communities are only the elderly, children under legal driving age, and older pre-retirement adults whose children have grown and left the family residence. The implications of this behavior pattern are ominous for sustainable community building.

The implications of creating "virtual communities" on the Internet are not yet clear to discern. Virtual communities, by definition, are "placeless" and not communities in the traditional sense. They are also anonymous in that interaction with other humans on the Web is impersonal and lacking in many important means of communication. Technology cannot yet deliver place complete with weather, smells, and animals, or the intimate human contact which community requires. Humans are drawn to Web sites out of common interests, or enclaves, and in that the individuals are not responsible for their actions, they are carnivalesque.

Interactions with other humans on the Web can ultimately have benefits if individuals are informed and motivated to join with their neighbors off-line and create their own real community. To settle for just a keyboard and screen is to only play the game. A good example of a Web-based network designed to cause real change is ***wholetruth.com***, a Web site designed to organize students from K-12 to learn about and fight the marketing of tobacco products to them. There are other Web-based, student-scientist networks that engage young people in active science activities with professional teachers and researchers.

Environment

The environment as a commons for all cultures, an overwhelmingly large, unspoiled, and inexhaustible resource, is an out-dated concept. Since the Apollo missions transmitted pictures of our fragile, watery planet back to Earth in the 1960's, our perspective of ourselves in relationship to our natural resources has continually changed towards one of collective responsibility both for our own welfare and for the welfare of all life on the planet. Economic value aside, there is no question that our lives are affected, and our lives affect, the quality of our oceans, atmosphere, fresh water bodies, soils, and all of the life they support.

The basic habitat requirements for all living things on this planet are to provide for: feeding, breeding, nesting, and resting. Certainly an over-simplification, these four activities and their requirements during the whole life cycle can define the basic needs for survival and sustained existence. For wild creatures, the habitat requirements to feed, breed, nest, and rest are supplied by natural environments and ecosystems that are relatively free of human presence and the effects caused by their inventions and activities.

Without these requirements being met, wildlife will migrate, if it can, to more favorable locations. This is not to say that humans and wild creatures cannot live in close proximity, they can. However, humans are too often ignorant of these four basic needs of wild creatures and they can intrude unnecessarily on their requirements. The lifestyles chosen by humans often introduce pollution, noise, and spatial conflict with local plants and animals. Many of these conflicts can be eliminated by proper planning and design that considers the needs of both the human and wildlife which is to live in the habitat. It is unfortunate that many pro-environmentalists have had to define their identity as anti-development because it has created a schism that proper design and planning may have avoided.

Even now, with lines in the sand already clearly drawn, it is possible to foresee bridges that could be built to span these different camps. The two factions have to discover their common ground and make a new sense of it that will benefit all parties. There are individuals already working on this important task and the slower-moving institutions are beginning to detect a change in the wind. The new common weal will grow from a new common will.

The recent revival of interest in traditional town planning and neighborhood design provides an opportunity to re-integrate native flora and natural wildlife into our communities through an understanding of their four basic needs and their habitat requirements. Initially driven by architects and urban designers, this renewed interest in neighborhood and town building is best exemplified by the Congress for New Urbanism (CNU), which has published its Charter and outlined its principles. Developed over a four-year period, CNU's Charter is now being scrutinized and expanded by many other disciplines and professions that can steer and propel urban development.

Introducing a strong environmental component into this Congress will enrich the debate and hopefully influence the direction CNU is promoting to include bio-diversity and a closer and more intimate living relationship of humans with local nature.

3. THE BUILT ENVIRONMENT BOTH SERVES HUMAN ACTIVITIES AND LIMITS THEM THROUGH STRUCTURAL CONSTRAINTS

The built environment—our towns and residences, our places of work, recreation and worship, as well as the roads, and transmission lines—is the theater in which we as individuals, and as a society, act out our lives. "All the world's a stage . . . ," said William Shakespeare and it is quite true, though not in the theatrical sense of a play that is separate from real life. Shakespeare also said, "The play's the thing . . ." and he meant that all life is for real.

Many civilizations have built large towns and cities that expressed their culture, fostered trade, and protected their lives and material wealth for centuries. However, archeological studies throughout much of the twentieth century show us that ancient civilizations have collapsed for ecological reasons. They exceeded their "carrying capacity," or their rates of use of energy and natural resources did not match the rates of supply and renewal.

This collapse of civilizations can be seen from the Fertile Crescent of the Middle East through to Meso-America and Asia. Mohenjo-daro, Angor Wat, Babylon, the early Mayans, Easter Island, pre-Hellenistic Greece, the dynasties of ancient Egypt, China, and even Rome all can trace this common pattern of excess, collapse, and slow retrenchment. The World's rate of consumption of energy and natural resources is *increasing* at an alarming rate as Asia and the Southern Hemisphere nations develop their petroleum-based economies. New concerns such as global warming, rising ocean levels, deforestation, desertification, and water shortages arise to dampen our optimism for a future without a collapse.

There have been and still are many cultures on earth that do not build extensive built environments requiring large infrastructures to support. The Aborigines of Australia and the Inuit of North America are examples of old cultures that have been sustainable for many centuries without large investments in built environments and infrastructures. There is also evidence in various locations on earth of cultures that did build large towns, cities, and supporting infrastructures and were not able to sustain themselves or their cultures. Their history and the reasons for their demise remain a partial mystery to us, but do not appear to be related to exceeding the local environmental carrying capacity. There are, however, other compelling reasons to explain why societies build.

According to an early 20th century biologist, A.J. Lotka, natural systems survive that can utilize available resources to build structures that capture those resources most effectively. Lotka's principle helps to explain succession in forestry, for example. The successive evolution of plant species and communities from pastureland, with high yield and low diversity, to a climax forest, with no yield and high diversity, is an exercise in superior plant genetics, applied serially over time. Each early succession plant species and associated insect and animal communities prepare the soils and immediate micro-climates for the next species. Early successional species grow and reproduce quickly and later ones more slowly; early species have low biomass and deteriorate quickly and later species have higher biomass and give up their greater resource investment to the detritus-eaters more grudgingly.

Looking at the built environments of human communities by analogy, early settlements build relatively quick, cheap buildings and settlements that survive to become towns and larger cities have greater investments in buildings and infrastructure, and so forth. The analogy between successional forestry and building human settlements seems to hold true until the climax state is reached. Climax means no more energy or resources are available to be used. A climax forest is processing every bit of available energy—sun, rain, wind—and utilizing all the minerals in the soil it can reach. It is producing no net yield year after year, just replacing what wears out and falls to the forest floor, blows away, or is consumed by other species. The forest stays the same by processing everything it can. It has reached the peak of its genetic capability. History teaches us that human societies go through these same patterns driven by the relationships among energy, populations, and resources. Our current concerns about sustainability are partly triggered by the realization that our nations are more and more closely linked and inter-dependent and are becoming a society that is global.

Individual nations are now positioning themselves to capture the necessary energy and resources that will enable them to maximize their cultural potential. Hopefully, these nations will realize the benefits of optimizing their relationships to other nations and the whole global society rather than contend with their neighbors for scarce resources. Our only real hope is to become a strong global community of nations working for the benefit of all.

The built environments of human communities have not yet reached the limit of the energy and resources they can acquire and process. Our Western civilization-based culture is very good at acquiring and exploiting resources. Refusing to acknowledge any limits, our economic engine continually locates new resources to sustain constant growth in building structure to find even more. If the limit for the forest is in the genetics of the most superior species to capture energy and resources, what is the limit for the collective activities of humans with their marvelous ingenuity and adaptability? Is this what happened to the civilizations whose empty, built environments we visit on vacations—did they outstripped their energy and resource base? Or were they genetically deficient and failed to adapt to their surrounding environment? Answers to these questions are elusive; and the important relationships between the built environment and the sustainability of the builders' community is unclear. That there is an important relationship is undeniable.

The physical permanence of the built environment ultimately becomes a limitation on how the community can adapt to necessary changes for its survival. The physical, built environment—the stage for acting out our collective lifestyles—limits our play. Unlike the morphological relationship between form and function found throughout nature, humans build their environments according to ideas they want to express in steel, wood, and stone. Form does not necessarily always follow function in the human built environment; form follows fiction. Even to the degree that environments built by humans represent mysterious relationships among the gods and cosmos, they still originate in the human imagination.

The relationship between the built environment and the potential quality of life of the members of the community is clearer to observe. At first, in the early stages of human settlement, the form of the built environment is arranged to suit the initial desired activities of the inhabitants, with the most basic needs addressed first. Feed, breed, nest, and rest become food, shelter, and physical protection and are taken care of first. Without those needs accomplished, nothing more sophisticated or esoteric can be easily accomplished. Once basic needs are satisfied, though, the human mind quickly begins to address its more esoteric and exotic interests. Religion, art, dance, warfare, mythology and storytelling, more elaborate forms of governance, and the necessity for public space and activities begin to emerge. So far, there is no need for large investments in built environments and supporting infrastructures to accommodate these new interests and house the emerging culture of the community. In fact, permanence may not ever be identified by the community as necessary. Desirable, perhaps; but not necessary. In an ironic contrast, some Buddhist cultures build elaborate and quite permanent facilities to institutionalize the contemplation of impermanence.

The increasing mobility among individuals and their families, and the fracturing of their relationships with their places and locales is in dichotomy with our large, permanent investments in built environments. The buildings, towns, and roads do not change very fast and do not adapt well at all compared to the change in our individual lifestyles and family living patterns. Our community-building capacity is diminished accordingly. Our built environment is becoming more of a commodity. We buy, sell and lease the use of it and move on.

Recent attempts to utilize technology as a surrogate for person-to-person community interaction has generally failed and the newest innovation, the Internet and World Wide Web, is failing also. Rather than establish communities with all the attributes of place, challenge of relationships, and capacity for adapting to necessary changes, the Net and WWW establish lifestyle enclaves of like-minded individuals who want to interact in a prescribed way with persons similar to themselves. No diversity, no sticking through tough times, and most importantly, no place to call home. World Wide Web surfers can log off and change their "virtual communities" at will; and with the anonymity of the Net, they can change their identities as well. Virtual communities are fabrications of the human mind and partial experiences at best, and misleading delusions at worst. Sure, the computer will eventually deliver bird calls, the sensation of mud on our feet, and probably a snakebite; but it can only be a surrogate for experiencing a real place, not a substitute for it.

In a similar vein, some global nations are expressing concerns for the loss of their history, heritage, and cultural identity due to the insatiable demands of U.S. entertainment companies. Disney, in particular, is singled out for creating entertainment from historical incidents at the expense of authenticity and integrity. As Neil Postman clearly stated with the title of a recent book, we are in danger of amusing ourselves to death by preferring everything to be entertaining.

For the built environment to contribute to a community's sustainability, it must exhibit the properties of adaptability, toughness and resiliency, diversity, energy efficiency and conservation of resources. To impose any small set of human contrived, rigid requirements or standards will only decrease its' usefulness to the resident community as it continually sustains itself.

4. FORMAL EDUCATIONAL PROCESSES MUST SUPPORT THE NON-FORMAL AND INFORMAL LEARNING THAT TAKES PLACE IN A COMMUNITY

The formal, public school system in the U.S. has evolved one-hundred-and-eighty degrees from its original purpose and intentions. This story is clearly written in the architectural history of school buildings. Setting aside for the moment private schools for the wealthy that were blatantly mimicking institutions in Europe, the one-room schoolhouse (note *house*) was a substitute for the original teaching space, the home living room. Its intention was to deliver an acute abbreviation of real life, epitomized by the three R's. It also was a more economical solution to the shortage of teachers, the wide dispersion of potential students, and the overworked and often illiterate parents. Accommodating students of all ages with direction from one generalist educator, the one-room schoolhouse was a social learning community organized for specific purposes within the larger community. The knowledge, skills and values taught there were a replica of the larger community, and under its direct supervision. Student attendance was irregular and sporadic with the seasons and weather, teachers were inconsistent.

Next came separate rooms for different age students and the definitions of "grades," the long road towards specialization had begun. It is questionable whether this was for the benefit of the students or the teachers. Now, we find ourselves with institutional buildings that house and represent the institution of public education, not our living rooms; and the process of education rarely creates a social learning community within the larger community, nor grows directly from it. In fact, the public schooling experience is disassociated from society to the degree that the present social learning community (for better or worse) occurs largely outside of the classroom and public school setting. The increasing violence in our schools, especially youth-on-youth crime, can be seen as bringing the media-influenced outside world into the schoolhouse instead of the learning process going the other way. School teachers and administrators desperately attempt to hold on to a 20th Century concept of public education against the onslaught of fractious parents and uncooperative students.

Public education will have to dramatically change in the 21st Century to serve its historic purpose in society, which, given the changes in our society already noted, may be hopelessly outdated. The constant refrain, from schoolteachers and administrators alike, for school reform will not correct this inversion of purposes. Even their success may be too little too late.

Young people have unprecedented access to information packaged for them. Much of it is designed and delivered to them by non-educators. Almost all of it is designed to be entertaining. They also have access to much of the same data that is used by governments and businesses in the daily workplace, which is available on the World Wide Web and Internet. The role for public education must evolve to serve young people in their struggle to grow and mature intellectually, morally, and spiritually by guiding their learning that occurs mostly outside the classroom.

One goal of economic interests and businesses is to push purchasing power lower and lower into younger age groups to increase markets for goods and services (mostly goods). A goal of government is to hold out tokens and privileges of maturity until later years. Legal voting, drinking, driving, and now smoking ages are argued to preserve a status quo which is continually being undercut by exposure to television, movies, and video games. The debate between social progression and academic preparation is beginning to rage, as the consequences both of these actions have on students are more clearly understood by educators and administrators alike.

Student educational experiences are no longer confined to the classroom, or guided only by professional teachers. The student's learning community is both impoverished at the local level and suffers from a surfeit of riches at the global level. Students amalgamate these experiences largely on their own according to their own interests, not according to historical epistemologies of knowledge. Students do not "read" a subject to acquire mastery and develop the capability to generate new knowledge for their chosen field of study. They accumulate disparate educational experiences, only a few of which come from formal public school, and assemble their own working knowledge of how things work and how to do the things they want to do.

In one sense, the public education system is evolving just as the successional forest does. Schooling in the U.S., early in the history of public education, was low diversity, high yield, and not very hardy or resistant to change. Assuming, for the sake of argument, that the public education system is now near a climax state; the characteristics it should be exhibiting are cooperation, collaboration, high rates of processing energy and resources, high diversity, adaptability, and stability in the face of changing outside forces. Clearly, our system of schooling is not near climax. A focus on sustainability education can help the public schooling system acquire those characteristics. Some schools are embracing this focus, others are poised for motion.

To be successful, public education has to dissolve its old identity as a separate, contained institution within the society and evolve into a new identity with branches and roots engaging the whole community. The physical structure of the schoolhouse must evolve to facilitate student learning within the whole community, not just the classroom. Educational experiences can be delivered where the students are, not just where the teachers, blackboards, and desks are. Educational experiences can be mobile, not just static with the occasional "field trips." The field can be the learning site.

The social skills developed in classroom schooling need to be preserved and enhanced to emphasize cooperation, collaboration, and team problem-solving. Inter-personal skills are also necessary to respect diversities of opinion, race, religion, and political affiliation. These are life skills that must be consistently reinforced both in schooling and the community at large.

Sustainability education has a rare target of opportunity where the latent messages of content, process, and delivery of educational experiences support and reinforce each other. In the words of Gregory Bateson, anthropologist, scholar, and student of learning, we must teach young people the "patterns which connect."

REFERENCES

1. *Illumination in the Flatwoods*; Joe Hutto; The Lions Press; 1995.
2. *Diffusion of Innovations*; Everett M. Rogers; The Free Press; New York; 1995.
3. *Topophilia*; Yi Fu Tuan.
4. *Hope, Human, and Wild*; Bill McKibben; Hungry Mind Press; Saint Paul, Minnesota; 1995.
5. *Geography of Nowhere*; William Howard Kunstler; Touchstone Books; 1994.
6. *Beyond Growth*; Herman Daly; Beacon Press; Boston, Massachusetts; 1996.
7. *Making Smart Growth Work in Florida: Part I-The Upward Spiral of a New Market; Part II-Making Smart Growth Easier; Part III-Making Conventional Development Pay Its Own Way*; Christopher Leinberger; Florida Sustainable Communities Center Website; <http://sustainable.state.fl.us>; posted May 1998.
8. *The Next American Metropolis: Ecology, Community, and the American Dream*; Peter Calthorpe; Princeton Architectural Press; Princeton, New Jersey; 1993.
9. *Visions for a New American Dream*; Anton Clarence Nelessen; Planners Press; Chicago, Illinois; 1993.
10. *Wired Neighborhood*; Stephen Doheny-Farina; Yale University Press; New Haven and London; 1996.
11. *Mind in Nature*; Gregory Bateson.

CHAPTER 15

General Motors Role in Education for Sustainability

Lori Wingerter

Education, both formal and lifelong, shapes an individual's predisposition and outlook on life, and it can therefore be a powerful tool to help ensure that sustainable practices become the norm by encouraging stakeholder action. Individuals should be taught how to become responsible in the use of resources, what sustainable consumption patterns are, and how their actions impact the environment. Effective sustainable practices must be incorporated into the education system in all relevant areas, including business. Education, therefore, is key to achieving a society that values and upholds sustainable practices.

At General Motors (GM), we have defined sustainable development as the process of seeking continuous improvement in our operations and products in a way that integrates economic, environmental, and social objectives into daily business decisions. This concept is particularly significant as we expand GM's global presence because it encompasses a responsible approach to growth and development—one in which the needs of today are satisfied without jeopardizing the needs of the future.

To ensure that sustainable practices become the norm, individuals must be educated about the interdisciplinary nature of sustainable systems,

Lori Wingerter, Coordinator, Community Impact Team, General Motors Corporation, Detroit, MI.

and be knowledgeable about how their decisions and actions impact the environment. The skills inherent to running a business, such as strategic planning and consensus building, are key to understanding and implementing sustainable practices. These skills are shared whenever businesses such as GM assist with community projects or enter into partnerships. Businesses can advance the assimilation of sustainable practices throughout society by providing technical assistance and mentoring opportunities. Therefore, we believe it is appropriate for business to take a leadership role in advancing education for sustainability.

GM's framework for education on sustainability consists of four components:

1) educating our employees; 2) extending this guidance to our suppliers, our dealers, and our customers; 3) entering into partnerships and collaborative initiatives with other stakeholders and outside institutions; and 4) engaging in dialogue with citizens, community organizations, and local government.

THE GM FRAMEWORK: THE FOUNDATION

According to GM, sustainable development is defined as an ongoing educational process of continuous improvement with discrete goals throughout. The baseline from which GM works regarding education for sustainability begins with the GM Environmental Principles (listed below). This set of six principles, together with a preamble that establishes their global applicability, was adopted by the General Motors Board of Directors in March, 1991. These principles have been communicated throughout the company as a framework for daily business conduct.

General Motors Environmental Principles

As a responsible corporate citizen, General Motors is dedicated to protecting human health, natural resources and the global environment. This dedication reaches further than compliance with the law to encompass the integration of sound environmental practices into our business decisions.

The following environmental principles provide guidance to General Motors personnel worldwide in the conduct of their daily business practices.

1. We are committed to actions to restore and preserve the environment.

2. We are committed to reducing waste and pollutants, conserving resources and recycling materials at every stage of the product life cycle.
3. We will continue to participate actively in educating the public regarding environmental conservation.
4. We will continue to pursue vigorously the development and implementation of technologies for minimizing pollutant emissions.
5. We will continue to work with all governmental entities for the development of technically sound and financially responsible environmental laws and regulations.
6. We will continually assess the impact of our plants and products on the environment and the communities in which we live and operate with a goal of continuous improvement.

THE GM FRAMEWORK: OUR EMPLOYEES—PRESENT AND FUTURE

Traditionally, and appropriately, large corporations like GM have focused their education efforts on their own employees. Consequently, GM's first avenue for integrating sustainability as a mindset begins with our employees. GM is committed to ensuring balanced economic development and heightened environmental and community awareness. This commitment, and the requisite education and training to put this commitment into practice, will be communicated to all employees through the General Motors University.

General Motors University (GMU) was created in 1997 to provide employees with the education and development needed to help GM succeed worldwide. Its focus is to bring about change through a network of learning opportunities designed to improve critical knowledge and skills. GMU will be instrumental in providing education about sustainability to our employees because it offers a mechanism for communicating common systems and best practices, and for enhancing individual competencies. GMU will advance the concept that environmental protection and economic objectives are interwoven in a common framework. This approach will be explained not only through traditional classroom learning, but through hands-on, functionally based training.

A group of employees are focused on activities that are strategically linked to sustainability: the GM environmental professional. Some of the courses offered through GMU will focus specifically on engineers and other plant personnel with daily environmental responsibilities. In keeping with the lifelong learning concept embraced by GMU, our environmental professionals participate in GM's annual environmental conference. These conferences have been held since 1993, and include presentations, training sessions, and workshops that foster the sharing of ideas, initiatives, and new learning about environmental stewardship. The conference offers an excellent opportunity to educate individuals about actions that will help ensure an environmentally sound and economically prosperous future.

GM recognizes the need to extend its educational efforts beyond the current workforce to include the workforce of tomorrow. GM supports an education system that promotes common themes and underlying principles rather than a specific program model for sustainability. We believe that this can best be accomplished by entering into partnerships with various higher education institutions. By playing a constructive role in external educational initiatives, we help to ensure that future employees will recognize the importance of factoring environmental implications into business decisions.

The GM Environmental Co-operative Education Program employs fourteen environmental engineering students from four universities: Michigan Technological University, Purdue University, Rensselaer Polytechnic University, and Wayne State University. The students' work assignments are rotated across many different manufacturing operations, which expose them to various situations in which the principles of sustainability are reinforced. Students who graduate are then placed into available openings at GM facilities.

GM established the GM Environmental Excellence Awards in the 1995–1996 school year. These awards are intended to recognize outstanding graduate students pursuing studies in environmental resource management or environmental economics, and allow GM to contribute to the development of future leaders in fields relating to the environment. The top priority in awarding these scholarships is to enable students to develop the skills necessary to evaluate national and global environmental issues.

The program combines academic studies with volunteer experiences addressing environmental issues. In addition to meeting the academic qualifications, applicants must submit a letter of recommendation from their school advisor and a personal commitment letter. In the personal commitment letter, students demonstrate their interest in pursuing a career in a field related to the environment, and their plan for completing the voluntary service requirement. This letter often is the decisive factor in selecting among the qualified applicants because it provides significant insight regarding the students' dedication.

In the U.S., a team that includes both GM and United Auto Workers (UAW) employees is developing training and reference materials for the GM Quality Network "Conserve Resources/Prevent Pollution (WE CARE)" Strategy. This is GM's pollution prevention strategy, which will be utilized worldwide. The WE CARE Action Strategy provides guidance to GM employees in reducing the impact of our operations and products on the natural environment. Each employee plays an important role in preventing or reducing pollution, recycling and reusing waste materials, and conserving natural resources on their job every day.

By using resources wisely, reducing the amount of waste generated, and better managing the wastes we do generate, substantial cost savings and benefits to the environment can be achieved. Further, many other benefits can result from a successful resource conservation and pollution prevention strategy. Other benefits include improved safety, better quality product, improved employee morale and public reputation, and reduced environmental liability.

THE GM FRAMEWORK: OUR SUPPLIERS, OUR DEALERS, OUR CUSTOMERS

GM believes that the depth and breadth of sustainable education can be strengthened by partnering with others. By working with groups with whom we share common interests, we can develop quality programs that promote sustainable systems.

Since 1994, GM has sponsored an annual *Green Day* to promote recycling, the use of materials that can be recycled and "design for the environment" awareness. The themes from prior years include *From Principles to Practice, From Practice to Profit,* and *Designing Today for Tomorrow's Environment.* Thousands of GM engineers and suppliers are briefed on recycling and "green design" successes that have been achieved by individual GM divisions and GM suppliers. This forum allows corporate management to demonstrate its support of each of the engineering groups while reinforcing the common theme. Supplier displays illustrate the influence of GM on the supply chain to encourage improved environmental performance, and to highlight parts and subsystems that have increased recyclability due to GM's initiatives.

Another very important GM initiative regarding sustainability is with our supplier community through the $PICOS_{TM}$ program. Established in 1989, this program sends teams of GM engineers to help suppliers identify faster, less costly, and more efficient ways of doing business. In 1996,

$PICOS_{TM}$ was expanded to promote environmental excellence by helping our suppliers identify and implement resource conservation and pollution-prevention (RCP2) projects that lower operating costs while reducing emissions. GM currently has over 200 supplier development engineers conducting over 2,000 workshops per year at supplier facilities to reduce waste and improve productivity. GM worked with the EPA, the DoE, Businesses for Social Responsibility, and Rutgers University to develop a training program for GM's supplier development engineers in the area of RCP2 to increase the environmental focus of these workshops. This training program had its worldwide inauguration in April, 1996. Through $PICOS_{TM}$, GM is promoting eco-efficiency in our supplier community.

The wide network of U.S. independent dealers selling GM products provides another important link to our framework on education for sustainability. GM provides each dealer an environmental resource guide which contains practical information regarding approaches that can improve environmental performance, and assists dealers regarding steps they can take to incorporate appropriate environmental actions into all of their business operations. The manual's four primary sections include: an audit booklet that enables dealers to conduct self audits of their facilities to assure compliance with environmental, as well as health and safety, regulations and recommended practices; a guidebook which provides practical guidelines and references to federal environmental regulations and key health and safety guidelines; the annual GM Environmental Report which provides an overview of the company's policies and practices; and a section listing "environmentally smart" equipment. This manual serves to educate GM dealers on ways their actions can protect the environment. All GM of Canada dealerships have also been provided reference manuals designed to assist their personnel to better understand the principles of waste management.

GM's customers and their children are also an important link in the framework. For example, GM has an educational partnership with the National SAFE KIDS® Campaign to promote the use and proper installation of child restraint systems in motor vehicles. The partnership will draw upon the Campaign's established network of more than 200 state and local coalitions in all 50 states, child care and health providers, and GM's network of dealers to disseminate key child safety messages.

THE GM FRAMEWORK: COLLABORATIVE INITIATIVES

Collaborations that reach beyond the traditional relationships with our employees, our suppliers, our dealers, and our customers serve as a catalyst for changing attitudes and behavior. These joint initiatives enable GM to address sustainability issues with a broader group of stakeholders, including other global businesses, environmental groups, and government agencies, and play an important role in assimilating sustainable practices throughout society.

An example of this type of cooperative initiative is the World Business Council for Sustainable Development (WBCSD). The WBCSD is a coalition of 125 international companies that are united by a shared commitment to the principles of sustainable development. GM's active involvement in the WBCSD and other business organizations is fundamental to our continued growth as a socially responsible corporation.

Similarly, GM's participation on the U.S. President's Council on Sustainable Development (PCSD) provides us the opportunity to learn from the broad views of the other business members, and representatives of environmental groups and governments. The PCSD was created by President Clinton in June, 1993, to advise him on ways to address the recognition that the quality of the country's future rests on integration of the economy, equity, and environment in national policy. In February 1996, this multistakeholder group reached consensus on an action plan (*Sustainable America: A New Consensus*) to achieve U.S. economic, environmental, and equity goals.

Perhaps one of the most important relationships for GM is with the Coalition for Environmentally Responsible Economies (CERES), a coalition of national environmental groups and socially responsible investors. In 1994, GM became the first major manufacturing company and the first Fortune 50 company to endorse the CERES Principles. CERES reciprocally endorsed the GM Environmental Principles as "consistent with the goals of the CERES Principles." Both principles share the objective of making environmental issues a part of all business decisions, and constitute a global commitment to protecting human health, natural resources, and the environment.

THE GM FRAMEWORK: COMMUNITY ENGAGEMENT

It is essential that education about sustainable development be extended to everyone—individuals, organizations, and communities. This educational effort begins by increasing citizen awareness and providing access to accurate information. By engaging in dialogue, stakeholders can share the knowledge needed to understand the interdependent relationship between the economy, the environment, and society. An informed public can find equitable solutions that are both economically and environmentally beneficial to all interested stakeholders.

Companies can involve the community by providing information about the environmental impact of their products and services. By publicizing their corporate environmental policies and practices, companies can educate the public and promote decisions that are more economically, environmentally, and socially responsible. Companies also can support the environmental education efforts of others. GM is actively involved in variety of these initiatives.

Since 1989, GM has worked in conjunction with local school systems worldwide to implement the Global Rivers Environmental Education Network (GREEN). Through GREEN, teachers all over the world are taking students to their local rivers and teaching them to monitor water quality, analyze watershed usage, identify the socioeconomic determinants of river degradation, and present their findings and recommendations to local officials.

In 1990, in cooperation with the U.S. Environmental Protection Agency, GM produced a children's educational video program on the environment called *I Need the Earth and the Earth Needs Me*. The 20-minute videotape is accompanied by a teacher's guide that suggests a variety of environmental education activities that focus on three basic areas of the environment—air, water, and soil. The activities fit into elementary-level curriculums—art, health, mathematics, music, physical education, and social studies, as well as science. To date, over 100,000 complimentary copies of the video tape and teacher's guide have been distributed worldwide. Language translations completed thus far include: Arabic, Chinese, French, German, Hebrew, Italian, Portuguese, Russian, and Spanish.

Community-based initiatives are another way to teach citizens about sustainability. Across the world, GM is working with community groups to plan for sustainability. GM and its employees are involved in various community programs aimed at improving environmental quality and education, such as:

- The McLaughlin Bay Wildlife Reserve is located on the north shore of Lake Ontario, behind GM of Canada's headquarters in Oshawa, Ontario. The 100-acre reserve is open to the public and is being developed cooperatively by GM of Canada and area environmentalists with the support of local government. Over three miles of walking trails for partially-sighted and sight-impaired hikers have been created and a long-term reforestation program initiated. Plans include transforming areas of the site into forest zones and shrub thickets to provide an environment suitable for nesting and breeding for resident and migrant birds and mammals.

- GM Powertrain facility in Bedford, Indiana has been active in local and regional environmental awareness. The plant hosted an Environmental Expo that included forty information booths highlighting topics on ways to minimize waste in homes and how to recycle. Local children were educated regarding the science of wastewater treatment. In addition, they visited a GM-sponsored exhibit entitled "EARTHQUEST: The Challenge Begins," which was housed at the Indianapolis Children's Museum. Plant employees have volunteered to clean up litter on a local roadway and educate youth about careers that are available in the science and engineering disciplines.
- GM Powertrain facility in Saginaw, Michigan hosted a UAW-GM Foundry Fun Fair that emphasized the importance of the environment. This event included demonstrations, and tours, and was geared to attract the youth of the community who were especially delighted with the four acres of wetlands that they toured.
- GM de Mexico participates in a community environmental education program that is implemented through conferences and video presentations in schools, social clubs, and other public places, radio and television programs, and national newspapers.
- Opel Taiwan created a program called "Initiatives for the Environment" to increase community awareness and promote environmental activism in Taiwan.

THE GM FRAMEWORK: LEARNING FROM THE PAST, BUILDING ON THE PRESENT, LOOKING TO THE FUTURE

Education for sustainability must start with our youth. Children need to learn how hard work, ingenuity, innovation, science, and technology can address environmental problems and prevent new ones from happening. Because it is easier to learn new approaches than to break old habits, it is essential that the youth of today be taught to make choices as responsible citizens of this global planet.

Business has an important role to play in education to increase the effectiveness of our schools in teaching the interrelationship of economic, environmental, and social objectives. Through education initiatives, we can ensure that all citizens become decision makers who understand the interdependence of economic prosperity, environmental protection, and social equity.

Education opportunities throughout people's lives enable them to adapt to changing economic and social conditions and respond to the need for environmental protection. All citizens must become decision makers who base their life style choices on their knowledge and understanding as responsible citizens of this planet. As business, we need to participate in a process that builds knowledge of the interdependence of economic prosperity, environmental protection, and social equity. Importantly for business, this is a 360° learning process. As we share knowledge with others, we gain knowledge in return.

CHAPTER 16

Partnerships for Progress
A Case Study from SC Johnson

Cynthia Georgeson

Four teens out shopping on a Saturday morning. Buying floor cleaner, air freshener, and laundry stain remover, the students make some environmentally conscious discoveries as they shop the aisles. What does this have to do with the success of a business at the dawn of the 21st Century? Everything! As we enter the new millennium, business success increasingly hinges on the ability of youth to understand and make decisions based on the fundamental concepts of sustainability.

Since 1990, industry leaders have invested significant resources to create processes that eliminate pollution and increase profitability. Many have identified tremendous revenue potential in the development of environmentally preferred methods to meet consumer needs. Yet the key to achieving environmental health and economic prosperity simultaneously remains dependent upon consumers with the knowledge to recognize and act upon environmental benefits. At high schools across the US, a global household products manufacturer hopes to raise awareness of the role of the individual consumer in influencing positive change by making more sustainable choices by teaching kids.

Cynthia Georgeson, Director Worldwide Communications, S.C. Johnson Corp., Racine, WI.

TEACHING KIDS TO SHOP SMART

The Saturday shopping excursion is the basis for a 15 minute educational video produced by students at John Marshall High School in Milwaukee, Wisconsin, through a unique school-to-work partnership between the inner city school and Racine, Wisconsin-based, family-owned manufacturer SC Johnson. The video "Get A Grip" is designed to stimulate interest in and an understanding of *eco-efficiency* in household consumer products as a significant step towards sustainable development. Eco-efficiency is an important component of industry efforts to operate sustainably by using fewer, more efficient materials to provide more performance with less overall waste, risk, and cost. The film's key message: consumers can encourage manufacturers to do more with less through their purchasing habits.

Considering factors such as package recyclability, product concentrates and reduced chemical usage, the teenage shoppers flash back to daily experiences, showing how eco-efficient decisions relate to their everyday lives. Students at John Marshall—using the hands-on experience to prepare for future careers in broadcasting—researched, wrote, produced, directed, and starred in "Get A Grip." The video, along with a companion teaching curriculum developed by *Keep America Beautiful*, has been sent to Natural and Social Science teachers at virtually every high school in the country. Because the film was created by students for students, SC Johnson believes it will engage the upcoming generation of household managers in more sustainable purchase habits. "Get A Grip" is the most recent in a series of environmental education videos and companion curricula sponsored by SC Johnson. Others include "Time & Time Again," a recycling and solid waste management teaching unit; "Ozone: The Hole Story," examining upper level ozone and causes for its depletion; and "Living Planet," a cinematic celebration of life on Earth.

CLOSING THE GAP BETWEEN CONCERN AND RESPONSIBLE ACTION

During the 1990s, SC Johnson commissioned a series of surveys by Roper Starch to gauge and correlate environmental knowledge, attitudes and action among North American adults and teens. The findings consistently revealed a high level of concern for the environment, but limited fact-based knowledge and even less individual action. Education for sustainability offers a bridge to close the gap between concern for Earth's future and the ability to act responsibly. That bridge is vitally important to the business community for a number of reasons.

First, industry efforts to produce sustainably must be matched by sustainable consumption patterns to maintain a healthy supply of natural resources necessary to do business and meet consumer needs in the future. Education for sustainability—which recognizes the inextricable link between a healthy environment and economy—promises a better-informed citizenry able to make decisions about complex environmental, economic, and social issues. Further, it assures a workforce capable and skilled in driving industry policies and practices towards ever more eco-efficient decisions.

Second, public perceptions of industry are sometimes formed on outdated or incorrect information while industry efforts to decrease environmental impact often go unrecognized. For example, since the mid-1970s household aerosol products have been free of chlorofluorocarbon propellants (CFCs) believed to deplete upper level ozone. In fact, SC Johnson led the world in unilaterally removing CFCs from its aerosol spray products worldwide in 1975. Three years later, the US government made the SC Johnson CFC-free aerosol standard law. Yet many people still think household aerosol products contain CFCs, a myth perpetuated in the media and in school textbooks. SC Johnson supports educational programs to clear up misperceptions and provide facts the public can use to evaluate the environmental impact of products they choose. An informed public capable of evaluating a company's or industry's balanced progress (or lack of progress) can influence more change for the better.

Industry actions to improve the environmental performance of products and processes alone cannot achieve global sustainability. Studies show people want to do the right thing when it comes to protecting the environment, but are largely unaware of their own unsustainable behaviors. Educational programs need to underscore the individual's role in protecting the environment. Through quality programs, students learn an awareness of the natural world; the ecological concepts which govern the natural and manmade environment; social, cultural, and economic factors which affect the environment; and the skills used by citizens to participate in the democratic process of evaluating and solving problems.

Just as corporations must ensure that employees are properly trained and have adequate resources to produce quality service or products, educational systems must provide teachers with the tools to promote quality education for a more sustainable world. Business can support education for sustainability through partnerships with local, state, and national efforts to improve environmental education in schools. This is one of many new roles for industry as a partner with communities working toward local and global sustainability.

REDEFINING THE ROLE OF BUSINESS IN SOCIETY

At the beginning of the 20th Century, at the height of the Industrial Age, corporate entities operated in relative autonomy from the communities around them—engines of economic growth with little involvement in other aspects of society. In the latter decades, particularly the 1970–80s, industry came under increased scrutiny for its contribution to and perceived lack of concern for societal problems, notably pollution and resource consumption. In the early 1990s, many businesses began putting in place measures to prevent pollution and conserve resources. Today, to meet raised expectations from stakeholders on many fronts, industry leaders are becoming much more active participants in their surrounding communities, involved in efforts to achieve broad-based social well-being—a key component of which is education.

For SC Johnson, commitment to community is not a new thing. In fact, since the company's founding in 1886, five generations of the Johnson family have held to the belief that communities should be better places for the company having been there. That commitment extends to employees throughout the world who participate actively in civic programs to enrich the quality of life locally, nationally, and globally. In recent years, corporate community involvement at SC Johnson has evolved in light of the company's long-term commitment to sustainable development. Today, the company's philanthropic giving and community relations programs are targeted directly at advancing the three legs of sustainability: economic vitality, environmental health, and social well-being—acknowledging that civic engagement is the key to success. Because sustainable communities are the foundation of a sustainable world, SC Johnson contributes to localized sustainability through civic leadership, employee volunteerism, targeted charitable contributions, and the responsible environmental management of operations.

BUILDING A SUSTAINABLE COMMUNITY

In Racine, Wisconsin, SC Johnson's world headquarters, the company has been a catalyst in dialogue and action as that community embarks on the path to sustainability. Providing leadership and financial resources, the company organized a Forum that was the first step in educating an entire community about the concept of sustainability and its importance to Racine's future, and the world. The Forum, held in July 1996, drew together a broad cross-section of community leaders and activists with national experts on sustainable development to begin an open discussion on community needs, and initiate a vision for meeting them. Not surprisingly, education was at the center of much of the discussions that day and in the days, months, and years to come.

Over the course of the next few months, the Sustainable Racine Initiative took shape. Nearly 1,500 dedicated individuals in the community identified a framework to engage all sectors of Racine in developing a vision for the community to pursue. Representatives from community groups, civic organizations, government bodies, the school district, churches and synagogues, and the business community pledged to work together to face Racine's challenges and apply resources toward a common vision. Today, through "shared-responsibility" partnerships, business, civic, and social organizations are working together for everyone's benefit.

Education is critically important to the community's vision for a brighter future with lasting progress dependent upon citizens with the ability to make sustainable decisions. In many ways, the quality of life in a community is tied to the effectiveness of the school system. Young people who are inspired by learning stay in school, get good jobs, give back to the community. More kids staying in school means less crime, less poverty, more people with the skills to do jobs well. All these factors contribute to a strong local economy, and make it in business' interest to support quality education—of which education for sustainability is a key part.

Partnering with communities for sustainability makes good business sense. Vital communities are more attractive to prospective employees; contribute to a happier, more productive workforce; and create a healthy atmosphere in which to conduct business and grow.

BUSINESS COMMUNITY MUST SUPPORT SUSTAINABLE EDUCATION

Education for sustainability is good for business. It assures a workforce with the knowledge and skills to compete effectively in a global marketplace increasingly tied to sustainable development. It prepares consumers to make informed decisions about products and product manufacturers while encouraging individuals to engage in sustainable behaviors. And it contributes to communities that are better able to meet the needs of all citizens—from individuals to corporations. Without a complete and balanced education on environmental, economic and social issues, citizens will be ill-prepared to face the challenges of a new century. Partnerships among the business community with local, state, and national educational efforts are needed to assure sustained progress.

SC Johnson is one of the world's leading manufacturers of specialty products for home storage, home and personal care, and insect control as well as products and services for commercial maintenance and industrial markets. A privately held, family-managed company, SC Johnson employs some 13,000 people in more than 50 countries around the world. The company is recognized globally for its heritage of leadership in strengthening communities and protecting the environment.

CHAPTER 17

Security and Sustainability
A Secure Nation in a Sustainable World

Michael Schneider

INTRODUCTION

Three decades ago, leading national security strategists developed an historic planning document, NSC 68. An aggressive communist Soviet Union and its neighbors presented the main danger to our vital national interests, and, the U.S. believed, to our survival. NSC 68 proposed a sweeping strategy "... to check and roll back the Kremlin's drive for world domination." Following George Kennan's historic "Mr. X" memo proposing containment of the USSR, and the Marshall Plan and Truman Doctrine, NSC 68 organized U.S. foreign policy for a generation. The document structured policies of nuclear deterrence and political, economic, and ideological competition in the pursuit of containment. The intensifying East-West rivalry influenced virtually all our foreign policy choices.

Thirty years later in the post-Cold War era, we need a new analysis of what are the primary threats, dangers and challenges to our interests, perhaps our survival. Are there opportunities for enhancing our interests?

Michael Schneider, Dean, Washington Program—Maxwell School of Business, Syracuse University, Washington, D.C.

What should the '90s counterpart to NSC 68 emphasize? Have the players and the rules changed? Are the changes so great that the game itself is different?

This essay tries to make three points: A broader definition of national security and more prominent attention to sustainability will advance U.S. strategic interests. The traditional instruments of national security have a vital role to play in assuring global sustainability. The nation needs to build its capacity to pursue sustainability through partnerships between government, business, and the non-profit sector.

THE CHANGING FACE OF WORLD AFFAIRS

In just a decade, international relations have changed dramatically. Instead of two super powers there is only the U.S., yet several other major states have the power and potential ability to influence events. The states of Europe move awkwardly toward confederation. Japan exerts great economic influence. Russia struggles with new institutions but remains one of two pre-eminent nuclear powers, and retains vast natural and human resources. China, India, and Brazil are also becoming multi-polar forces despite current and future setbacks. Other states such as Korea, Malaysia, Thailand, South Africa, Nigeria, Mexico will exercise greater influence. Regional groupings—North America, the European Union, APEC, ASEAN, the Mercosur states—are shaping the global economy on a broader level.

Economic globalization is changing the way we work, conduct business, consume goods, pursue happiness, identify with our community and nation and the larger world. Financial transactions occur instantly, 24 hours-a-day; surges and declines in markets in Tokyo, Hong Kong, Bangkok and further west create waves across every time zone every day. Instead of trades that required months or weeks or even days, companies can move goods and services around the globe in hours and shift work to contractors and labor in distant places with ease. This speed, complexity, and flexibility appears to create greater independence but also reflect a more interdependent system. What happens in the oil fields of Saudi Arabia, the farms of Guatemala, the data processing centers in Ireland or India and thousands of sites around the globe affects American citizens at home or at work all of the time.

Increasingly, it is also becoming difficult to identify corporations by nationality. Mergers and acquisitions meld former national icons. Multinational corporations are the engines of global growth. These huge conglomerates are the equals of nation-states in wealth, and they are revising the rules of engagement and challenging national sovereignty in significant ways.

In particular, knowledge-based business, information, and telecommunications are reformulating the emerging international order. Computers and high-speed communications are speeding the flow of trade and finance. Connectivity is unprecedented, spreading and accelerating as analog gives way to digital communication. Innovation is expanding the knowledge base dramatically as well. For each leap in speed and processing capacity there is a new way to catalogue, store and retrieve—and use—the new information. Yet it is unclear whether wise choices are any more possible, or our security and stability any more likely now or in the long run. Information overload poses one of the truly new challenges to international order and daily life.

Publics count much more than ever before in history. National decision making is no longer the province of only a small group of elites. Leaders seek to appeal to, mobilize, and respond to their publics. As democracy takes root and elections become more frequent, fair and free, and as greater transparency develops, the rules of governance open up the game to many new players. "Non-state actors" and "non-governmental organizations" are textbook symbols of the dramatic growth in participation in national and international life. Even the remaining authoritarian regimes can not entirely ignore their own public or world opinion.

The new NSC 68 for the early 21st century, therefore, must deal with a vastly different world arena, with more actors, more complex interactive issues, much more communication and a new public dimension independent of, yet affecting, government. While unresolved or un-attended issues of the past half-century form the bases for the next, America's security will rely on a new vision and a new strategy that addresses both "old" and "new" challenges to global sustainability.

MAJOR GLOBAL ISSUES FOR THE ADVENT OF THE NEW CENTURY

The Cold War left a residue of unsolved or unrecognized challenges. While the world was preoccupied with the Cold War, many issues grew quietly for the past several decades and blossomed in the immediate aftermath of the dissolution of the Warsaw Pact and the movement toward free market democracy around the globe. The Cold War also either suppressed or exacerbated conflicts that have flared up anew in the past decade. East-West rivalry left vast numbers of nuclear, bio-chemical, and conventional weapons, and many nations with the ability to build them. Thus we face "old" or continuing issues, "new" concerns, and a number of "old-new" issues, rooted historically but appearing to be novel.

The challenges before us involve many nations across regional lines. Although current issues and their solutions will be particular to each country, there are common elements. There are "traditional security" dimensions and elements relating to sustainability in all the cases. No single nation will solve these issues on its own, nor in most cases can even a regional grouping act on its own successfully. None of the crises are purely economic, military, political, or humanitarian. Successful outcomes will require cooperation among many nations, multi-national business, non-governmental and international organizations. Finally, U.S. leadership will be essential.

Regional Conflicts

Whether suppressed or exaggerated by the Cold War, conflicts in the Middle East, Bosnia, the Indo-Pakistan subcontinent, the Horn of Africa, the Caucuses pose the central dangers for world peace as we enter the new millennium. Each current or potential crisis could draw in U.S. forces and assets, with great costs of many kinds.

While requiring crisis responses at many junctures, none of these conflicts will finally be resolved without a sustained combination of approaches. We will need diplomacy, the implied or actual use of military force, and surely development assistance and private investment as well as novel institutional approaches to ameliorate the conditions and underlying causes of the conflicts. Most of all, the world community will need time to chip away at these impacted problems—which goes against our own national desire for quick answers.

Global Financial Instability

It is unclear how rapidly the financial crisis in Asia can be resolved, nor whether other countries such as Russia, South Africa, Brazil will deal effectively with intensifying problems of financial mismanagement. In each case, albeit with somewhat different contexts, these are political and cultural as well as economic crises, and surely would bring down regimes and threaten systems if conditions worsen. These non-technical/economic dimensions merit greater recognition in both analysis and response.

Proliferation of Weapons of Mass Destruction

Although U.S. national security strategy required a series of alliances and military preparation to contain Soviet expansion, the victory has its costs today. The U.S. and Russia retain huge nuclear arsenals. The recent Indo-Pakistani nuclear tests point to the continuing dangers of nuclear proliferation. Chemical and biological weapons pose horrific dangers. Unprecedented proliferation of conventional arms pose equally significant, if different, dangers to global stability. A new strategy to contain the spread of weapons is needed, yet competes with the perceived economic and defense interests of the producing and consuming states.

Narcotics Trafficking and Abuse

No issue better exemplifies the false dichotomy between "domestic" and "foreign affairs." Substance abuse and trafficking is probably the major single contributor to crime and violence, social disorganization, reduced productivity and significant health care costs in the U.S. Military and police approaches to the problem, while essential, have had mixed success at best. Long-term approaches such as alternative development assistance in producing countries, and demand reduction activities domestically merit even greater emphasis

Terrorism

Feeding on social and political instabilities, economic disparities, and longstanding ethnic and communal antipathies, global terrorism, whether state supported or home grown, will not be eliminated absent a combination of police, intelligence, military and legal approaches *plus* new ways of understanding and co-opting or ameliorating the contributory factors.

Climate Change

While appearing to pit environmentalists against business interests, the resolution of this issue will require close cooperation and lead to a new era of invention, energy development and consumption, transportation and industrial production. Absent action by the U.S., the major industrializing nations will not be induced to move away from fossil fuels, but without their commitments, the U.S. Congress is not likely to move forward in the near future. Even if the debate in the U.S. produces a stalemate in the next few years, multi-national business is beginning to see the opportunities and will therefore make the changes that help reduce greenhouse gases. The big crapshoot is whether science and technology, economics, and politics will come into conjunction soon enough to mitigate the effects of global warming.

Loss of Natural Resources

Related to global warming, deforestation, loss of arable land, marine fish stocks, and biodiversity are having multiple impacts on international affairs. It shouldn't take too much scrutiny to see the connections between these changes and civil war, regional conflicts, and calls for international intervention resulting from the conflicts over land and resources or other instabilities.

Growing International Public Health Dangers

The growing AIDS pandemic in Africa and Asia shows few signs of abating. Also possible are outbreaks of disease or even pandemics on a scale of the 1918 influenza outbreak with huge losses of life, breakdowns in public health and economic and political destabilization. There is a war underway between microbes and humanity every bit as serious as the various wars we have fought in the past century. The recent U.S. initiative against infectious diseases points out the important role of U.S. military research, logistics and field operations to the success of international civilian public health efforts.

Population Growth and Global Demographic Shifts

Underlying all of the challenges we must confront is the huge growth in global population of the past half century. The impacts and implications are almost incalculable. The debate in the U.S. over abortion has had a chilling effect on our support for programs that would stabilize the rate of population growth. There can be little doubt of the multiple impacts on global health, the stability of nations, the utilization of natural resources by the momentum that will increase world population from 5.5 to 8 billion people in the next 25 years. Humanitarian crises, regional conflicts, deforestation, over-fishing and cultivation of marginal land, spread of diseases, massive movements of people will increase in part because of population growth. If the United States and the world community can't deal with the underlying cause, the crisis responses will be all the more costly and ineffective.

Each of these global challenges has a distinct impact on our daily lives. For the first time perhaps we can not live securely at home, without addressing both at home and abroad the range of issues beyond our borders. This fact calls into question the compartmentalization of U.S. foreign policy, military-intelligence, and domestic programs and budget allocations.

CURRENT U.S. STRATEGY

The Clinton Administration's 1997 report, "A National Security Strategy for a New Century," recognizes the nation's assets in the new era: ". . . Our military might is unparalleled; a dynamic global economy offers increasing opportunities for American jobs and American investment; and the community of democratic nations is growing, enhancing the prospects for political stability, peaceful conflict resolution and greater hope for the people of the world."

The report also takes serious note of unresolved problems: "At the same time, the dangers we face are unprecedented in their complexity. Ethnic conflict and outlaw states threaten regional stability; terrorism, drugs, organized crime and proliferation of weapons of mass destruction are global concerns that transcend national borders; and environmental damage and rapid population growth undermine economic prosperity and political stability in many countries." Necessarily so broad a report tends to compartmentalize issues. Each priority is treated independently, with some effort to make connections to a larger vision. Nor is it clear whether the Clinton Administration will carry out in the same balanced way the policies emphasized by the National Security Strategy Review.

Secretary of State Madeleine Albright put the sustainability issue more personally in a 1998 Earth Day speech: "The threats we face from environmental harm are not as spectacular as those of a terrorist's bomb or missile. But we know that the health of our families will be affected by the health of the global environment. The prosperity of our families will be affected by whether other nations develop in sustainable ways. The safety of our families will be affected by whether we cut back on the use of toxic chemicals. And the security of our nation will be affected by whether we are able to prevent conflicts from arising over scarce resources." Secretary Albright picked up and expanded upon two rather slender policy threads of her predecessor, Warren Christopher, who belatedly sought to move sustainability concerns from the back to the front burner.

WHY SUSTAINABILITY?

How can a concept as apparently vague as "sustainability" or "sustainable development" serve as a centralizing vision for the future? After all, national security for so long has rested on military power, implied or used, and on economic power, political will and national unity (or internal control.) However, traditional emphases of security don't adequately address all of the issues that are likely to confront us in the coming century. The old repertoire doesn't give us the analytic skills and understanding, even the proper time frame and benchmarks to preserve and promote our broad interests. We are left chasing headlines, coping with crises or violent upheavals, essentially trying—with mixed results—to remediate the past more than shape the future. The issues that have emerged in full public view in the past decade require longer time, more cooperation within and among more nations around the globe, and a greater blend of politics, economics, social action, and community involvement. "Domestic" and "international" programs will require greater integration.

The concept of sustainability helps provide the understanding and the capacity to deal strategically with the wider range of cross-cutting global issues that increasingly dominate international affairs, from arms proliferation, terrorism, and drugs, to climate change or population growth. This concept calls for innovation, participation, decentralization to serve an essentially conservative notion of incremental, ameliorative change and protection of freedom and human dignity.

The concept of strategic sustainability is the successor to the bipolar emphases of the Cold War era. Sustainability will require in real terms a mix of democratic governance, transparency and a strong civil society, well-managed market economies and the wise application of science and technology. The concept is based on a series of continua rather than the classic zero-sum game, on multipolarity rather than bipolarity, on degrees of success or failure rather than survival or devastation. Sustainability involves a new set of strategic objectives that give greater consideration to the "new" issues as well as the carry-over from the Cold War, and to the public dimension of national and international affairs.

TOWARD A BROADER DEFINITION OF NATIONAL SECURITY

This list of contemporary challenges and the governing concept of sustainability suggest the need for a national security strategy based on a broader definition of "security." U.S. foreign and security policy needs to look beyond the current crises and traditional modes of analyzing and dealing with issues and the behavior of other nations. Mechanisms, roles, priorities of the Cold War era need revision if we are to lead the world successfully into the next century. In a far more complex world, security relies on public health as well as military strength, on maintaining forests well as building homes.

Three distinct points of view, or groups of advocates for different policy priorities and emphases, are apparent: traditional foreign and economic policy and military communities ("geostrategists" and "global marketers") and newer internationally active environmental, health, and social issues communities ("sustainability advocates.") The three groupings have genuine differences, made worse by inexperience with each other. To some extent, traditionalists see the newcomers as overly idealistic, naïve about the use and misuse of power, the ambitions and evils of rogue states and leadership. In the extreme, sustainability advocates see traditionalists as narrow minded, focused on the same old power game, or on shortsighted and exploitative economic gain.

The traditionalist hierarchy of threats and dangers starts first with immediate military or paramilitary dangers to our very survival and broadens to long-term corrosive dangers or challenges. Geostrategists would protect the state and its populace, and maintain military and related economic power to use if necessary, but preferably to deter aggression if possible. The nation-state and its instruments of power are central. Economic strength, political stability and social coherence within the nation-state are fundamental. Strong security and political alliances are vital. In the Cold War framework, the battles were ideological, political, economic; all issues, trends, new developments were evaluated through the prism of their implications for our competitive standing with our rivals.

The post-Cold War era brings fewer immediate threats of total annihilation or of long-term ideological-political-economic losses. Nevertheless, training, experience, and current responsibilities lead many traditionalists to focus more on the current dangers than on the more indirect pressures arising from long-term forces such as increasing population and declining carrying capacity.

Global marketers point to globalization as the central organizing force for the post-Cold War era. Narrowly perceived, in only its economic dimension, globalization from the marketers' standpoint is a powerful integrating force in which the market places a premium on innovation, flexibility and speed. Corporations have the ability to shift resources and business, call upon labor supplies globally. The rising tide is global not merely local or regional. The global market must still preserve stability and predictably, but by adopting new rules and adjusting to the new transactional technologies.

Geostrategists ask where does the nation-state and sovereignty come into this equation, and what about important political and security controls over business. The sustainability advocates express alarm about the impact of economic globalization on the environment and resources, on social well being, community control and individual autonomy. The marketers answer the geostrategist concern by emphasizing that globalization indeed calls for new rules of engagement among states and international organizations, but that market forces will create long term stability by raising income levels across the globe and requiring stable legal systems. Sustainability needs will be met as the market finds new technologies and business opportunities that rely less on polluting or environmentally destructive means.

The geostrategists, global marketers, and sustainability advocates will choose different priorities from the list of issues outlined above. The first will continue to pursue the military balance of power, with added interest in counter narcotics and crime and terrorism. The second group will place higher emphasis on U.S. economic and financial, market and trade concerns, while the third will stress the "new" global social, environmental and health needs and opportunities. However our national interests and world leadership will benefit from the three groups seeking common ground, reinforcement and cooperation, and mitigating differences where possible.

THE NEED TO SHAPE A SUSTAINABLE STRATEGY FOR NATIONAL SECURITY

In what ways can sustainability advocates, geostrategists, and global marketers work together? Where there are differences, are these irreconcilable? Is there common ground or opportunity for compromise or more sympathetic understanding? What steps are needed?

Understand Different Perspectives

First, on the all-important definition of national interests or national security: all parties need to stretch themselves to understand the others' vantage and seek more common ground. National interests have always included cross-regional goals—energy security, open sea-lanes, freedom from attack. These need to be broadened, beyond traditional bilateral or geopolitical relations. A sustained dialogue is needed among the three groupings of interests and opinion over the relative importance and priority among our national interests in stopping the flow of narcotics, in reducing greenhouse gases, in contributing to the stabilization of the rate of population growth, or reducing the conditions that produce massive and disruptive refugee or migration movements.

Broaden the Concept of National Security

Similarly the concept of "national security" should be broadened to include our prevention or response to other threats than traditional military or major economic concerns such as denial of energy in a crisis. The range is great—from paramilitary to counter-terrorist action, to steps that slow the proliferation of weapons of mass destruction. Public health needs to be considered an element of national security. Therefore we can not be secure if flooding or frequent storms occur as a result of global warming and agriculture is affected, or if a new infectious disease threatens to invade the United States. And would we remain unaffected if the Indo-Pakistan rivalry escalates into major conventional, or even nuclear, war, or brutal conflict resumes in Rwanda and Burundi or Haiti? If the economies of Asia, including Japan, continue to turn down, will Wall Street and American investors, continue to prosper, untouched?

The so-called "new issues" remind the policy community, that there is much more to national security and national interest than the immediate or imminent crisis, regime failure, military action, or other acute challenge to stability.

The new agenda also reminds us of the connections among forces and issues, and the need to consider the impact and implications of this overlap. Can we achieve peace and security in the Middle East without dealing with water rights and sharing? Can the cycles of violence in Rwanda and Burundi cease without steps to address the history of tribal antipathies and the pressures placed on warring parties from the lack of adequate land? Or can the production of cocaine cease without alternative opportunities for the farmers?

Fortunately many in government realize the connections among issues, and among different dimensions of U.S. engagement in world affairs. The National Security Strategy reviews of the 1990s go well beyond traditional defense plans.

The Pentagon has a phrase for the change—"Operations Other than War." Much effort, training and analysis are spent in preparing U.S. military for very different responsibilities in coming decades. No other institution in the U.S. is so well prepared and has the capacity to respond to natural disasters, humanitarian crises requiring rapid delivery of supplies and short-term technical assistance. And U.S. military forces have played an absolutely critical role in international peacekeeping efforts. The Defense Mapping Agency, and U.S. military satellite systems are major sources of space-based information vitally needed for a range of developmental and environmental concerns. The clean energy, environmental remediation and planning activities of the Department are critical. U.S. military forces are a major global consumer of energy. Care, commitment, technical capacity and planning are needed to clean up our numerous military installations here and abroad and prevent environmental damage. Thus the U.S. military is a critically important partner in U.S. sustainability efforts. This must be more broadly recognized, and even broader efforts made by sustainability advocates to reach out to our military.

Better Integrate and Coordinate "Domestic" and "International" Programs

More cooperation by domestic and international agencies in the executive branch or by the separate oversight committees on Capitol Hill might help the nation deal with issues that are both domestic and international. A new cooperative team approach has been tried regarding a few issues such as counter-narcotic action and is needed across the board in the executive branch to deal with strategic concerns. NIDA, Justice, the DEA, HHS, State and AID work together on narcotics problems, as does CDC, PHS and DoD on infectious diseases here and abroad. The same cross-sectoral cooperation in all phases of policy is needed for all of the global concerns cited above.

THE NEED FOR NEW, CROSS-CUTTING STRATEGIES

The United States has tended to develop strategies along traditional political-economic lines or in bilateral terms. These have served us well in the past, but there is a need for an equally compelling and consistently implemented set of international strategies based on the broad concept of sustainability, dealing with global issues. For example:

Long-term Strategy for Energy Independence/New Forms of Energy

For several years, oil has been exceedingly inexpensive, and just when we begin to believe there may be a shortage some day, new sources are discovered or new technologies developed that make oil recovery possible and comparatively inexpensive. But our good fortune won't last forever. With rapid growth predictable around the world, the boom times for cheap oil can not be assumed. The United States and other industrialized nations are dependent on imported petroleum, and will pay a heavy price if the supply is reduced, or cut off or increasing demand leads to increased prices.

What about the health costs of rapidly increasing use of fossil fuels, especially coal, and oil? Will the move toward natural gas be fast enough to offset the pollution and global warming impacts of using ever-increasing amounts of coal and oil? Conversely, are there not important market opportunities for technologies to clean up and use fossil fuels more efficiently? And what about the political costs of relying on external supply of oil and natural gas?

Clearly, the United States and other nations need a comprehensive long-term strategy to move away from coal and oil, to a mix of energy sources. A move to natural gas for a period of time will be determined by market interests, governmental policies and technological innovations, and should be followed by a transition to renewable energy sources. This strategy needs definition, will be adjusted as innovation and consuming and living patterns change, as economic growth occurs and demographic shifts take place. It should have a prominent place among our national and international goals. We should continuously examine our consumption and figure out how to improve our use of energy. This will have important ramifications on a host of geopolitical, economic and social concerns for the next century.

Conflict Resolution and Water Resources is a second area in which the United States, with other governments and international organizations, development institutions and NGOs, needs a long term strategic approach. Water issues play a different role in each region, yet few would gainsay the vital importance of water to the region—its development, public health, life or death. If water is the cause or contributor to conflicts, it is also possible to conceive of cooperation, confidence building and conflict resolution efforts built around water sharing, preservation or development.

Promotion of U.S. Environmental Technologies presents a third area in which a long-term effort, undertaken by the private sector with incentives and facilitation by national, state and local governments will advance both U.S. trade/investment and sustainability interests. This has been recognized in part by U.S. initiatives such as the Asian-American Environmental Initiative of the Bush Administration and the Clinton Administration Trade Promotion Coordinating Council and special efforts, under the International Trade Administration at Commerce to advance the sale of such technologies, products and services.

Support for Transparency and the Civil Society is required for the development of healthy democracy, thriving market economies and sustainable societies. The United States Government and diverse non-governmental institutions have striven for decades to advance the rule of law and democracy around the globe. Here too, a more sustained effort with greater policy, financial and institutional support is needed. We have a patchwork effort, and need to better integrate, or at least relate, efforts among federal international affairs and development agencies, and generate a broader commitment to support non-governmental programs. We are witnessing proliferating investment and development schemes, NGO ties and links at all levels of society via the Internet, but the mutual reinforcement of political, economic, and environmental goals is somewhat absent.

Periodic professions of support for this sector by recent Administrations are not fulfilled by comparable funding commitments and follow up.

A vital subset of our interest in promoting democratization includes the promotion of free, open, and modern communications systems. This of course has enormous economic consequences for the U.S. and developing nations. Nowhere is our stake in development more clear than in the information and telecommunications sector. The implications for sustainability are equally significant. Societies will use modern, high-speed telecommunications systems to share knowledge and ideas, build cooperation and coalitions that lead to greater openness, responsible and responsive government. Modernized telecommunications and information systems will help communities make decisions that advance the commonweal, from health, to environmental protection, to quality of life. Much of this is occurring through self-initiated market forces, or in the non-market sector, through transnational cooperation of NGOs. Yet a national policy commitment that brings together the private sector, the non-governmental organizations, city, state, and federal agencies would help the long-term effort become less scattered, identify special needs, such as regional connectivity or national training.

A National Commitment to Broader and Stronger International Educational Cooperation is imperative but missing. Although the federal government spends annually some $1.6 billion on international education, exchange and training, there are gaps in performance and program coordination. The Fulbright Program, the flagship exchange and international education endeavor of the United States, has been taken for granted by Administrations and cut severely in the past decade by Congress. Fulbright does not have the assets required to fulfill our interests effectively in a more diverse world. Our global leadership depends on greater international education for American young people, yet we lag in foreign language and area studies. Although more students study and travel abroad their experience is superficial. While the number of foreign students in the U.S. has grown to some 450,000 in the past year—and produces some $7 billion in income for the U.S.—the nation needs to continue to attract foreign talent. This is not a task only for the federal government, but federal funds are needed to match and foster other commitments. States and localities are aggressively expanding their ties with counterparts abroad and these efforts almost always involve educational exchange and cooperation. Yet without a broad and coherent support strategy from federal agencies, the effort will be fragmentary and important opportunities will be missed.

For example, the President's Council for Sustainable Development prepared a marvelous set of documents laying out concepts and practical steps, sources of counsel, cooperation, and support for international education. But the distribution and use of these documents and the exploitation of this dimension of the entire enterprise has languished for lack of policy support by the U.S. Government or of commanding involvement from one or more of the leading non-governmental organizations. Highly dedicated professionals in the field of sustainability education are thwarted by a combination of lack of funding, lack of a clear and strong policy mandate, and lack of opportunity to work more consistently together. Yet education will establish the infrastructure of understanding and commitment, and build the constituencies around the globe for greater emphasis on the elements of sustainability as part of national and international priorities.

PRACTICAL STEPS TO BUILD THE CAPACITY FOR SUSTAINABILITY

Special efforts are needed to meld the expertise and vantages of the three broad communities into a cohesive global strategy. Specifically:

- The upgrading of our national intelligence and research capacities with regard to global issues and the many elements of sustainability should continue. The National Intelligence Council, military intelligence, State Department's fledgling office of the Geographer and of Global Issues should all expand their mandate and gain more tools of research. More resources are needed for the development of GIS applications for important U.S. policy concerns. The example of the use of space based imagery to bring warring factions in the Dayton accords together is but one instance of the potential of GIS for sustainability and U.S. interests. Other examples would include various maritime disputes, including fishery areas and migration patterns, or water resource issues, land/boundary disputes and regional planning, etc. The potential uses are high of GIS for the full range of U.S. engagements.
- Similarly new approaches to sustainability research and educational exchange should be the focus of a coordinated effort by NSF, AID, EPA, State, Energy, USIA, and others.
- Training programs and educational opportunities for U.S. foreign service, military and civil service personnel in foreign affairs and security agencies should be examined with an eye to introduction of sustainability concepts and much more integrated treatment of the issues and accompanying approaches outlined above.

- The Federal government should provide greater incentives for professional development and advancement through work on global issues. The traditional national security and foreign policy agencies especially have much work to do to revise their reward and promotion systems to account more for those who work on sustainability concerns.
- A major effort is needed to develop an integrated strategy for U.S. Government support for Internet access and use to foster sustainable development. This is not a substitute but an incentive for market expansion in developing countries, and will coincidentally advance U.S. business interests in technology transfer and in supplying the infrastructure of communication and information flows in the next century.
- Active cooperation in conflict prevention and resolution, in peacekeeping and in transitional justice is imperative, but requires both consistent, ongoing federal involvement and leadership, and also the forging of long-term partnerships with multilateral organizations and transnational NGOs. There is growing recognition in the foreign affairs, military and legal communities of the advantages of innovative approaches in this area, but also a need for a fresh analysis of opportunities and limitations, particularly of the role of military forces.

A SECURE NATION IN A SUSTAINABLE WORLD

Sustainability depends on security, but security equally depends on sustainability. The two concerns are no longer separable, if they ever were. Geostrategists, global marketers and sustainability advocates share more common interests than they may recognize. Priorities will differ, most notably in matching resources to issues and in choosing what concerns to address first. Surely there are genuine differences such as approaches to trade issues that make labor, human rights or environmental considerations secondary or perceptions of how markets work or ought to work. Some political-military or intelligence priorities will undoubtedly offend sustainability advocates, and some initiatives on behalf of environmental protection or social policies will clash with bilateral or geostrategic requirements. But if we elevate and broaden our international engagement, and place strategy over tactics and long-term over immediate gains, the three communities will work together more than they ever believed possible.

About the Editors

Keith Wheeler is the Executive Director of the Center for a Sustainable Future, a division of the Concord Consortium, an educational and technology research and development institute located in Shelburne, Vermont. The Center for a Sustainable Future specializes in the integration of science and technology in creating new learning tools about sustainability education, worldwide. Keith was the first Executive Director and CEO of Global Rivers Environmental Education Network (GREEN), a 135 nation international non-governmental organization that was a recognized leader in watershed education and conservation. He was the Assistant Director for the Adirondack Park Agency, leading the development of Interpretative and Environmental Education Centers in New York State. Keith served as a soil scientist and international resource development specialist for the USDA and for Cornell University from 1976–1988.

Keith has served on a number of commissions and task forces including: The President's Council for Sustainable Development (PCSD), Public Linkage and Education Task Force; Co-Chair of the White House led initiative "Education for Sustainability"; Co-Chair for the National Forum for Partnerships Supporting Education about the Environment; Commissioner for the International Union for Conservation and Nature (IUCN)–CEC; a member of the Conservation Fund's Conservation Leadership Project; a member of Vermont 2020; a trustee to the Brandwein Institute; a trustee for the Pocono Environmental Education Center (PEEC); a trustee for Earth Force; a trustee for the White Mountain School; an advisor to the General Motors Corporation for Sustainability; a member of the Capital Campaign committee for Shelburne Farms; and the Founding Chairman for the Empire State Pedologists. Keith has taught several courses at Cornell University. He has presented numerous keynote addresses to business, scientific, technologic and education conferences throughout the world on a variety of environmental and educational policy issues.

Anne Peracca Bijur co-directs Vermont's Building Education for a Sustainable Society project, a professional development program concerning education for sustainability for K–12 grade teachers. She formerly coordinated a successful effort to integrate sustainability concepts into Vermont's Framework of Standards and Learning Opportunities. She is a graduate of Cornell University and a candidate for a master's degree from the University of Vermont.

Index

Printed in the United Kingdom
by Lightning Source UK Ltd.
119261UK00006B/73

9 780306 464201